Honda H100 H100S Singles Owners Workshop Manual

by Jeremy Churchill
with an additional Chapter on the H100 S II model
by Pete Shoemark

Models covered
H100 A. 99cc. February 1980 to February 1984
H100 S-D. 99cc. June 1983 to September 1986
H100 S-G (H100 S II). 99cc. March 1986 to January 1988
H100 S-J (H100 S II). 99cc. January 1988 to February 1992

ISBN 978 1 85010 877 1

Printed in England *(734-7R9)*

ABC

Haynes Publishing Group
Sparkford Nr Yeovil
Somerset BA22 7JJ England

Haynes Publications, Inc
859 Lawrence Drive
Newbury Park
California 91320 USA

Acknowledgements

Our thanks are due to Paul Branson Motorcycles of Yeovil who supplied the H100 A and H100 S-G models and to Frank's Autoneeds of Langport who supplied the H100 S-D model. The H100 S-J featured on the front cover was supplied by Fran Ridewood & Co. of Wells.

We would like to thank the Avon Rubber Company, who kindly supplied information and technical assistance on tyre fitting; NGK Spark Plugs (UK) Ltd for information on sparking plug maintenance and electrode conditions, and Renold Ltd for advice on chain care and renewal.

About this manual

The purpose of this manual is to present the owner with a concise and graphic guide which will enable him to tackle any operation from basic routine maintenance to a major overhaul. It has been assumed that any work would be undertaken without the luxury of a well-equipped workshop and a range of manufacturer's service tools.

To this end, the machine featured in the manual was stripped and rebuilt in our own workshop, by a team comprising a mechanic, a photographer and the author. The resulting photographic sequence depicts events as they took place, the hands shown being those of the author and the mechanic.

The use of specialised, and expensive, service tools was avoided unless their use was considered to be essential due to risk of breakage or injury. There is usually some way of improvising a method of removing a stubborn component, providing that a suitable degree of care is exercised.

The author learnt his motorcycle mechanics over a number of years, faced with the same difficulties and using similar facilities to those encountered by most owners. It is hoped that this practical experience can be passed on through the pages of this manual.

Where possible, a well-used example of the machine is chosen for the workshop project, as this highlights any areas which might be particularly prone to giving rise to problems. In this way, any such difficulties are encountered and resolved before the text is written, and the techniques used to deal with them can be incorporated in the relevant section. Armed with a working knowledge of the machine, the author undertakes a considerable amount of research in order that the maximum amount of data can be included in the manual.

A comprehensive section, preceding the main part of the manual, describes procedures for carrying out the routine maintenance of the machine at intervals of time and mileage. This section is included particularly for those owners who wish to ensure the efficient day-to-day running of their motorcycle, but who choose not to undertake overhaul or renovation work.

Each Chapter is divided into numbered sections. Within these sections are numbered paragraphs. Cross reference throughout the manual is quite straightforward and logical. When reference is made 'See Section 6.10' it means Section 6, paragraph 10 in the same Chapter. If another Chapter were intended, the reference would read, for example, 'See Chapter 2, Section 6.10'. All the photographs are captioned with a section/paragraph number to which they refer and are relevant to the Chapter text adjacent.

Figures (usually line illustrations) appear in a logical but numerical order, within a given Chapter. Fig. 1.1 therefore refers to the first figure in Chapter 1.

Left-hand and right-hand descriptions of the machines and their components refer to the left and right of a given machine when the rider is seated normally.

Motorcycle manufacturers continually make changes to specifications and recommendations, and these, when notified, are incorporated into our manuals at the earliest opportunity.

We take great pride in the accuracy of information given in this manual, but motorcycle manufacturers make alterations and design changes during the production run of a particular motorcycle of which they do not inform us. No liability can be accepted by the authors or publishers for loss, damage or injury caused by any errors in, or omissions from, the information given.

Contents

Left-hand view of Honda H100 A

Right-hand view of Honda H100 S

Honda H100 A engine close-up

Introduction to the Honda H100

The present Honda empire, which started in a wooden shack in 1947, now occupies vast factory space, covering all aspects of modern motorcycle design, research, testing and production. The facilities available to the large work force are second to none and the model range covers every conceivable aspect of motorcycling.

Although better known for their use of small single-cylinder four-stroke engines in the machines that form their comprehensive range of commuter machines, Honda expanded this range still further with the introduction of the H100–A to the UK in February of 1980. Its single-cylinder two stroke engine is designed to attract the rider who prefers the basic simplicity and economy of this configuration.

The H100-A is designed to fulfil the role of an economical and practical commuter machine. Features such as CDI ignition, the separate oil injection system for engine lubrication, and the fully-enclosed rear chain combine to make it a simple machine to operate in everyday use, requiring the bare minimum of attention. The 100cc engine gives adequate performance for use in normal traffic conditions, but without the high fuel consumption of more highly-tuned two-stroke machines. Engine efficiency is enhanced by the use of a reed valve induction system and the vibration inherent in a single-cylinder engine is largely cancelled out by the fitting of a single-shaft primary balancer.

The H100 S-D was introduced in June of 1983 and was sold initially alongside the H100-A model which continued unchanged until stocks were exhausted in 1983/84. It is a heavily revised version of the H100 which has lost some of its more practical features as a result of the restyling operation which was felt necessary to give it wider appeal. Although very similar in design and specification, the new model differs in many details; for example the oil tank is positioned behind the right-hand side panel, a lightweight chromed top cover replaces the full-enclosure final drive chain casing, a tachometer is fitted and, somewhat unusually, the CDI ignition is replaced by a contact breaker system.

The H100 S-D was eventually superseded by the H100 S II during 1986. Reference should be made to Chapter 7 for information relating to the H100 S II model.

Dimensions and weight

	H100	H100 S
Overall length	1885 mm (74.2 in)	1860 mm (73.2 in)
Overall width	780 mm (30.7 in)	690 mm (27.2 in)
Overall height	1040 mm (40.9 in)	1020 mm (40.2 in)
Wheelbase	1225 mm (48.2 in)	1200 mm (47.3 in)
Ground clearance	160 mm (6.3 in)	160 mm (6.3 in)
Dry weight	86 kg (190 lb)	83.5 kg (184 lb)

Ordering spare parts

When ordering spare parts for the model featured in this manual it is advisable to deal direct with an official Honda agent, who will be able to supply many of the items required ex-stock. It is advisable to get acquainted with the local Honda agent, and to rely on his advice when purchasing spares. He is in a better position to specify exactly the parts required and to identify the relevant spare part numbers so that there is less chance of the wrong part being supplied by the manufacturer due to a vague or incomplete description.

When ordering spares, always quote the frame and engine numbers in full, together with any prefixes or suffixes in the form of letters. The frame number is found stamped on the right-hand side of the steering head, in line with the forks. The engine number is stamped on the left-hand side of the crankcase, immediately in front of the gearchange lever shaft.

Use only parts of genuine Honda manufacture. A few pattern parts are available, sometimes at cheaper prices, but there is no guarantee that they will give such good service as the originals they replace. Retain any worn or broken parts until the replacements have been obtained; they are sometimes needed as a pattern to help identify the correct replacement when design changes have been made during a production run.

Some of the more expendable parts such as sparking plugs, bulbs, tyres, oils and greases etc., can be obtained from accessory shops and motor factors, who have convenient opening hours, and can often be found not far from home. It is also possible to obtain parts on a Mail Order basis from a number of specialists who advertise regularly in the motorcycle magazines.

Location of frame number

Location of engine number

Safety first!

Professional motor mechanics are trained in safe working procedures. However enthusiastic you may be about getting on with the job in hand, do take the time to ensure that your safety is not put at risk. A moment's lack of attention can result in an accident, as can failure to observe certain elementary precautions.

There will always be new ways of having accidents, and the following points do not pretend to be a comprehensive list of all dangers; they are intended rather to make you aware of the risks and to encourage a safety-conscious approach to all work you carry out on your vehicle.

Essential DOs and DON'Ts

DON'T start the engine without first ascertaining that the transmission is in neutral.

DON'T suddenly remove the filler cap from a hot cooling system – cover it with a cloth and release the pressure gradually first, or you may get scalded by escaping coolant.

DON'T attempt to drain oil until you are sure it has cooled sufficiently to avoid scalding you.

DON'T grasp any part of the engine, exhaust or silencer without first ascertaining that it is sufficiently cool to avoid burning you.

DON'T allow brake fluid or antifreeze to contact the machine's paintwork or plastic components.

DON'T syphon toxic liquids such as fuel, brake fluid or antifreeze by mouth, or allow them to remain on your skin.

DON'T inhale dust – it may be injurious to health (see *Asbestos* heading).

DON'T allow any spilt oil or grease to remain on the floor – wipe it up straight away, before someone slips on it.

DON'T use ill-fitting spanners or other tools which may slip and cause injury.

DON'T attempt to lift a heavy component which may be beyond your capability – get assistance.

DON'T rush to finish a job, or take unverified short cuts.

DON'T allow children or animals in or around an unattended vehicle.

DON'T inflate a tyre to a pressure above the recommended maximum. Apart from overstressing the carcase and wheel rim, in extreme cases the tyre may blow off forcibly.

DO ensure that the machine is supported securely at all times. This is especially important when the machine is blocked up to aid wheel or fork removal.

DO take care when attempting to slacken a stubborn nut or bolt. It is generally better to pull on a spanner, rather than push, so that if slippage occurs you fall away from the machine rather than on to it.

DO wear eye protection when using power tools such as drill, sander, bench grinder etc.

DO use a barrier cream on your hands prior to undertaking dirty jobs – it will protect your skin from infection as well as making the dirt easier to remove afterwards; but make sure your hands aren't left slippery. Note that long-term contact with used engine oil can be a health hazard.

DO keep loose clothing (cuffs, tie etc) and long hair well out of the way of moving mechanical parts.

DO remove rings, wristwatch etc, before working on the vehicle – especially the electrical system.

DO keep your work area tidy – it is only too easy to fall over articles left lying around.

DO exercise caution when compressing springs for removal or installation. Ensure that the tension is applied and released in a controlled manner, using suitable tools which preclude the possibility of the spring escaping violently.

DO ensure that any lifting tackle used has a safe working load rating adequate for the job.

DO get someone to check periodically that all is well, when working alone on the vehicle.

DO carry out work in a logical sequence and check that everything is correctly assembled and tightened afterwards.

DO remember that your vehicle's safety affects that of yourself and others. If in doubt on any point, get specialist advice.

IF, in spite of following these precautions, you are unfortunate enough to injure yourself, seek medical attention as soon as possible.

Asbestos

Certain friction, insulating, sealing, and other products – such as brake linings, clutch linings, gaskets, etc – contain asbestos. *Extreme care must be taken to avoid inhalation of dust from such products since it is hazardous to health.* If in doubt, assume that they *do* contain asbestos.

Fire

Remember at all times that petrol (gasoline) is highly flammable. Never smoke, or have any kind of naked flame around, when working on the vehicle. But the risk does not end there – a spark caused by an electrical short-circuit, by two metal surfaces contacting each other, by careless use of tools, or even by static electricity built up in your body under certain conditions, can ignite petrol vapour, which in a confined space is highly explosive.

Always disconnect the battery earth (ground) terminal before working on any part of the fuel or electrical system, and never risk spilling fuel on to a hot engine or exhaust.

It is recommended that a fire extinguisher of a type suitable for fuel and electrical fires is kept handy in the garage or workplace at all times. Never try to extinguish a fuel or electrical fire with water.

Note: *Any reference to a 'torch' appearing in this manual should always be taken to mean a hand-held battery-operated electric lamp or flashlight. It does **not** mean a welding/gas torch or blowlamp.*

Fumes

Certain fumes are highly toxic and can quickly cause unconsciousness and even death if inhaled to any extent. Petrol (gasoline) vapour comes into this category, as do the vapours from certain solvents such as trichloroethylene. Any draining or pouring of such volatile fluids should be done in a well ventilated area.

When using cleaning fluids and solvents, read the instructions carefully. Never use materials from unmarked containers – they may give off poisonous vapours.

Never run the engine of a motor vehicle in an enclosed space such as a garage. Exhaust fumes contain carbon monoxide which is extremely poisonous; if you need to run the engine, always do so in the open air or at least have the rear of the vehicle outside the workplace.

The battery

Never cause a spark, or allow a naked light, near the vehicle's battery. It will normally be giving off a certain amount of hydrogen gas, which is highly explosive.

Always disconnect the battery earth (ground) terminal before working on the fuel or electrical systems.

If possible, loosen the filler plugs or cover when charging the battery from an external source. Do not charge at an excessive rate or the battery may burst.

Take care when topping up and when carrying the battery. The acid electrolyte, even when diluted, is very corrosive and should not be allowed to contact the eyes or skin.

If you ever need to prepare electrolyte yourself, always add the acid slowly to the water, and never the other way round. Protect against splashes by wearing rubber gloves and goggles.

Mains electricity and electrical equipment

When using an electric power tool, inspection light etc, always ensure that the appliance is correctly connected to its plug and that, where necessary, it is properly earthed (grounded). Do not use such appliances in damp conditions and, again, beware of creating a spark or applying excessive heat in the vicinity of fuel or fuel vapour. Also ensure that the appliances meet the relevant national safety standards.

Ignition HT voltage

A severe electric shock can result from touching certain parts of the ignition system, such as the HT leads, when the engine is running or being cranked, particularly if components are damp or the insulation is defective. Where an electronic ignition system is fitted, the HT voltage is much higher and could prove fatal.

Working conditions and tools

When a major overhaul is contemplated, it is important that a clean, well-lit working space is available, equipped with a workbench and vice, and with space for laying out or storing the dismantled assemblies in an orderly manner where they are unlikely to be disturbed. The use of a good workshop will give the satisfaction of work done in comfort and without haste, where there is little chance of the machine being dismantled and reassembled in anything other than clean surroundings. Unfortunately, these ideal working conditions are not always practicable and under these latter circumstances when improvisation is called for, extra care and time will be needed.

The other essential requirement is a comprehensive set of good quality tools. Quality is of prime importance since cheap tools will prove expensive in the long run if they slip or break when in use, causing personal injury or expensive damage to the component being worked on. A good quality tool will last a long time, and more than justify the cost.

For practically all tools, a tool factor is the best source since he will have a very comprehensive range compared with the average garage or accessory shop. Having said that, accessory shops often offer excellent quality tools at discount prices, so it pays to shop around. There are plenty of tools around at reasonable prices, but always aim to purchase items which meet the relevant national safety standards. If in doubt, seek the advice of the shop proprietor or manager before making a purchase.

The basis of any tool kit is a set of open-ended spanners, which can be used on almost any part of the machine to which there is reasonable access. A set of ring spanners makes a useful addition, since they can be used on nuts that are very tight or where access is restricted. Where the cost has to be kept within reasonable bounds, a compromise can be effected with a set of combination spanners – open-ended at one end and having a ring of the same size on the other end. Socket spanners may also be considered a good investment, a basic $3/8$ in or $1/2$ in drive kit comprising a ratchet handle and a small number of socket heads, if money is limited. Additional sockets can be purchased, as and when they are required. Provided they are slim in profile, sockets will reach nuts or bolts that are deeply recessed. When purchasing spanners of any kind, make sure the correct size standard is purchased. Almost all machines manufactured outside the UK and the USA have metric nuts and bolts, whilst those produced in Britain have BSF or BSW sizes. The standard used in USA is AF, which is also found on some of the later British machines. Others tools that should be included in the kit are a range of crosshead screwdrivers, a pair of pliers and a hammer.

When considering the purchase of tools, it should be remembered that by carrying out the work oneself, a large proportion of the normal repair cost, made up by labour charges, will be saved. The economy made on even a minor overhaul will go a long way towards the improvement of a toolkit.

In addition to the basic tool kit, certain additional tools can prove invaluable when they are close to hand, to help speed up a multitude of repetitive jobs. For example, an impact screwdriver will ease the removal of screws that have been tightened by a similar tool, during assembly, without a risk of damaging the screw heads. And, of course, it can be used again to retighten the screws, to ensure an oil or airtight seal results. Circlip pliers have their uses too, since gear pinions, shafts and similar components are frequently retained by circlips that are not too easily displaced by a screwdriver. There are two types of circlip pliers, one for internal and one for external circlips. They may also have straight or right-angled jaws.

One of the most useful of all tools is the torque wrench, a form of spanner that can be adjusted to slip when a measured amount of force is applied to any bolt or nut. Torque wrench settings are given in almost every modern workshop or service manual, where the extent to which a complex component, such as a cylinder head, can be tightened without fear of distortion or leakage. The tightening of bearing caps is yet another example. Overtightening will stretch or even break bolts, necessitating extra work to extract the broken portions.

As may be expected, the more sophisticated the machine, the greater is the number of tools likely to be required if it is to be kept in first class condition by the home mechanic. Unfortunately there are certain jobs which cannot be accomplished successfully without the correct equipment and although there is invariably a specialist who will undertake the work for a fee, the home mechanic will have to dig more deeply in his pocket for the purchase of similar equipment if he does not wish to employ the services of others. Here a word of caution is necessary, since some of these jobs are best left to the expert. Although an electrical multimeter of the AVO type will prove helpful in tracing electrical faults, in inexperienced hands it may irrevocably damage some of the electrical components if a test current is passed through them in the wrong direction. This can apply to the synchronisation of twin or multiple carburettors too, where a certain amount of expertise is needed when setting them up with vacuum gauges. These are, however, exceptions. Some instruments, such as a strobe lamp, are virtually essential when checking the timing of a machine powered by CDI ignition system. In short, do not purchase any of these special items unless you have the experience to use them correctly.

Although this manual shows how components can be removed and replaced without the use of special service tools (unless absolutely essential), it is worthwhile giving consideration to the purchase of the more commonly used tools if the machine is regarded as a long term purchase Whilst the alternative methods suggested will remove and replace parts without risk of damage, the use of the special tools recommended and sold by the manufacturer will invariably save time.

Fault diagnosis

Contents

1 Introduction

This Section provides an easy reference-guide to the more common ailments that are likely to afflict your machine. Obviously, the opportunities are almost limitless for faults to occur as a result of obscure failures, and to try and cover all eventualities would require a book. Indeed, a number have been written on the subject.

Successful fault diagnosis is not a mysterious 'black art' but the application of a bit of knowledge combined with a systematic and logical approach to the problem. Approach any fault diagnosis by first accurately identifying the symptom and then checking through the list of possible causes, starting with the simplest or most obvious and progressing in stages to the most complex. Take nothing for granted, but above all apply liberal quantities of common sense.

The main symptom of a fault is given in the text as a major heading below which are listed, as Section headings, the various systems or areas which may contain the fault. Details of each possible cause for a fault and the remedial action to be taken are given, in brief, in the paragraphs below each Section heading. Further information should be sought in the relevant Chapter.

Engine does not start when turned over

2 No fuel flow to carburettor

● Fuel tank empty or level too low. Check that the tap is turned to 'On' or 'Reserve' position as required. If in doubt, prise off the fuel feed pipe at the carburettor end and check that fuel runs from pipe when the tap is turned on.
● Tank filler cap vent obstructed. This can prevent fuel from flowing into the carburettor float bowl bcause air cannot enter the fuel tank to replace it. The problem is more likely to appear when the machine is being ridden. Check by listening close to the filler cap and releasing it. A hissing noise indicates that a blockage is present. Remove the cap and clear the vent hole with wire or by using an air line from the inside of the cap.
● Fuel tap or filter blocked. Blockage may be due to accumulation of rust or paint flakes from the tank's inner surface or of foreign matter from contaminated fuel. Remove the tap and clean it and the filter. Look also for water droplets in the fuel.
● Fuel line blocked. Blockage of the fuel line is more likely to result from a kink in the line rather than the accumulation of debris.

3 Fuel not reaching cylinder

● Float chamber not filling. Caused by float needle or floats sticking in up position. This may occur after the machine has been left standing for an extended length of time allowing the fuel to evaporate. When this occurs a gummy residue is often left which hardens to a varnish-like substance. This condition may be worsened by corrosion and crystalline deposits produced prior to the total evaporation of contaminated fuel. Sticking of the float needle may also be caused by wear. In any case removal of the float chamber will be necessary for inspection and cleaning.
● Blockage in starting circuit, slow running circuit or jets. Blockage of these items may be attributable to debris from the fuel tank by-passing the filter system or to gumming up as described in paragraph 1. Water droplets in the fuel will also block jets and passages. The carburettor should be dismantled for cleaning.
● Fuel level too low. The fuel level in the float chamber is controlled by float height. The fuel level may increase with wear or damage but will never reduce, thus a low fuel level is an inherent rather than developing condition. Check the float height, renewing the float or needle if required.

4 Engine flooding

● Float valve needle worn or stuck open. A piece of rust or other debris can prevent correct seating of the needle against the valve seat thereby permitting an uncontrolled flow of fuel. Similarly, a worn needle or needle seat will prevent valve closure. Dismantle the carburettor float bowl for cleaning and, if necessary, renewal of the worn components.
● Fuel level too high. The fuel level is controlled by the float height which may increase due to wear of the float needle, pivot pin or operating tang. Check the float height, and make any necessary adjustments. A leaking float will cause an increase in fuel level, and thus should be renewed.
● Cold starting mechanism. Check the choke (starter mechanism) for correct operation. If the mechanism jams in the 'On' position subsequent starting of a hot engine will be difficult.
● Blocked air filter. A badly restricted air filter will cause flooding. Check the filter and clean or renew as required. A collapsed inlet hose will have a similar effect. Check that the air filter inlet has not become blocked by a rag or similar item.

5 No spark at plug

● Ignition switch not on.
● Fuse blown.
● Spark plug dirty, oiled or 'whiskered'. Because the induction mixture of a two-stroke engine is inclined to be of a rather oily nature it is comparatively easy to foul the plug electrodes, especially where there have been repeated attempts to start the engine. A machine used for short journeys will be more prone to fouling because the engine may never reach full operating temperature, and the deposits will not burn off. On rare occasions a change of plug grade may be required but the advice of a dealer should be sought before making such a change. 'Whiskering' is a comparatively rare occurrence on modern machines but may be encountered where pre-mixed petrol and oil (petroil) lubrication is employed. An electrode deposit in the form of a barely visible filament across the plug electrodes can short circuit the plug and prevent its sparking. On all two-stroke machines it is a sound precaution to carry a new spare spark plug for substitution in the event of fouling problems.
● Spark plug failure. Clean the spark plug thoroughly and reset the electrode gap. Refer to the spark plug section and the colour condition guide in Chapter 3. If the spark plug shorts internally or has sustained visible damage to the electrodes, core or ceramic insulator it should be renewed. On rare occasions a plug that appears to spark vigorously will fail to do so when refitted to the engine and subjected to the compression pressure in the cylinder.
● Spark plug cap or high tension (HT) lead faulty. Check condition and security. Replace if deterioration is evident. On rare occasions the cap may break down, thus preventing sparking. If this is suspected, fit a new cap as a precaution.
● Spark plug cap loose. Check that the spark plug cap fits securely over the plug and, where fitted, the screwed terminal on the plug end is secure.
● Shorting due to moisture. Certain parts of the ignition system are susceptible to shorting when the machine is ridden or parked in wet weather. Check particularly the area from the spark plug cap back to the ignition coil. A water dispersant spray may be used to dry out waterlogged components. Recurrence of the problem can be prevented by using an ignition sealant spray after drying out and cleaning.

● Ignition switch shorted. May be caused by water corrosion or wear. Water dispersant and contact cleaning sprays may be used. If this fails to overcome the problem dismantling and visual inspection of the switch will be required.

● Shorting or open circuit in wiring. Failure in any wire connecting any of the ignition components will cause ignition malfunction. Check also that all connections are clean, dry and tight.

● Ignition coil failure. Check the coil, referring to Chapter 3.

● Capacitor (condenser) failure (H100 S only). The capacitor may be checked most easily by substitution with a replacement item. Blackened contact breaker points indicate capacitor malfunction but this may not always occur.

● Contact breaker points pitted, burned or closed up (H100 S only). Check the contact breaker points, referring to Routine maintenance. Check also that the low tension leads at the contact breaker are secure and not shorting out.

● Faulty electronic ignition system components (H100 only). Refer to Chapter 3.

● Faulty diodes (H100 S only). Refer to Chapter 3.

6 Weak spark at plug

● Feeble sparking at the plug may be caused by any of the faults mentioned in the preceding Section other than those items in the first three paragraphs. Check first the contact breaker assembly (H100 S only) and the spark plug, these being the most likely culprits.

7 Compression low

● Spark plug loose. This will be self-evident on inspection, and may be accompanied by a hissing noise when the engine is turned over. Remove the plug and check that the threads in the cylinder head are not damaged. Check also that the plug sealing washer is in good condition.

● Cylinder head gasket leaking. This condition is often accompanied by a high pitched squeak from around the cylinder head and oil loss, and may be caused by insufficiently tightened cylinder head fasteners, a warped cylinder head or mechanical failure of the gasket material. Re-torqueing the fasteners to the correct specification may seal the leak in some instances but if damage has occurred this course of action will provide, at best, only a temporary cure.

● Low crankcase compression. This can be caused by worn main bearings and seals and will upset the incoming fuel/air mixture. A good seal in these areas is essential on any two-stroke engine.

● Piston rings sticking or broken. Sticking of the piston rings may be caused by seizure due to lack of lubrication or overheating as a result of poor carburation or incorrect fuel type. Gumming of the rings may result from lack of use, or carbon deposits in the ring grooves. Broken rings result from over-revving, over-heating or general wear. In either case a top-end overhaul will be required.

Engine stalls after starting

8 General causes

● Improper cold start mechanism operation. Check that the operating control functions smoothly. A cold engine may not require application of an enriched mixture to start initially but may baulk without choke once firing. Likewise a hot engine may start with an enriched mixture but will stop almost immediately if the choke is inadvertently in operation.

● Ignition malfunction. See Section 9. Weak spark at plug.

● Carburettor incorrectly adjusted. Maladjustment of the mixture strength or idle speed may cause the engine to stop immediately after starting. See Chapter 2.

● Fuel contamination. Check for filter blockage by debris or water which reduces, but does not completely stop, fuel flow, or blockage of the slow speed circuit in the carburettor by the same agents. If water is present it can often be seen as droplets in the bottom of the float bowl. Clean the filter and, where water is in evidence, drain and flush the fuel tank and float bowl.

● Intake air leak. Check for security of the carburettor mounting and hose connections, and for cracks or splits in the hoses. Check also that the carburettor top is secure.

● Air filter blocked or omitted. A blocked filter will cause an over-rich mixture; the omission of a filter will cause an excessively weak mixture. Both conditions will have a detrimental effect on carburation. Clean or renew the filter as necessary.

● Fuel filler cap air vent blocked. Usually caused by dirt or water. Clean the vent orifice.

● Choked exhaust system. Caused by excessive carbon build-up in the system, particularly around the silencer baffles. Refer to Chapter 2 for further information.

● Excessive carbon build-up in the engine. This can result from failure to decarbonise the engine at the specified interval or through excessive oil consumption. Check the oil pump adjustment.

Poor running at idle and low speed

9 Weak spark at plug or erratic firing

● Battery voltage low. In certain conditions low battery charge, especially when coupled with a badly sulphated battery, may result in misfiring. If the battery is in good general condition it should be recharged; an old battery suffering from sulphated plates should be renewed.

● Spark plug fouled, faulty or incorrectly adjusted. See Section 5 or refer to Chapter 3.

● Spark plug cap or high tension lead shorting. Check the condition of both these items ensuring that they are in good condition and dry and that the cap is fitted correctly.

● Spark plug type incorrect. Fit plug of correct type and heat range as given in Specifications. In certain conditions a plug of hotter or colder type may be required for normal running.

● Contact breaker points pitted, burned or closed-up (H100 S only). Check the contact breaker assembly, referring to Routine maintenance.

● Ignition timing incorrect.

● Faulty ignition coil. Partial failure of the coil internal insulation will diminish the performance of the coil. No repair is possible, a new component must be fitted.

● Faulty capacitor (condenser) – H100 S only. A failure of the capacitor will cause blackening of the contact breaker point faces and will allow excessive sparking at the points. A faulty capacitor may best be checked by substitution of a serviceable replacement item.

● Defective flywheel generator ignition source. Refer to Chapter 3 for further details on test procedures.

● Faulty diodes (H100 S only). Refer to Chapter 3.

10 Fuel/air mixture incorrect

● Intake air leak. Check carburettor mountings and air cleaner hoses for security and signs of splitting. Ensure that carburettor top is tight.

● Mixture strength incorrect. Adjust slow running mixture strength using pilot adjustment screw.

● Pilot jet or slow running circuit blocked. The carburettor should be removed and dismantled for thorough cleaning. Blow

through all jets and air passages with compressed air to clear obstructions.
● Air cleaner clogged or omitted. Clean or fit air cleaner element as necessary. Check also that the element and air filter cover are correctly seated.
● Cold start mechanism in operation. Check that the choke has not been left on inadvertently and the operation is correct.
● Fuel level too high or too low. Check the float height, renewing float or needle if required. See Section 3 or 4.
● Fuel tank air vent obstructed. Obstructions usually caused by dirt or water. Clean vent orifice.

11 Compression low

● See Section 7.

Acceleration poor

12 General causes

● All items as for previous Section.
● Choked air filter. Failure to keep the air filter element clean will allow the build-up of dirt with proportional loss of performance. In extreme cases of neglect acceleration will suffer.
● Choked exhaust system. This can result from failure to remove accumulations of carbon from the silencer baffles at the prescribed intervals. The increased back pressure will make the machine noticeably sluggish. Refer to Chapter 2 for further information on decarbonisation.
● Excessive carbon build-up in the engine. This can result from failure to decarbonise the engine at the specified interval or through excessive oil consumption. Check oil pump adjustment.
● Ignition timing incorrect (H100S). Check the contact breaker gap and set within the prescribed range ensuring that the ignition timing is correct. If the contact breaker assembly is worn it may prove impossible to get the gap and timing settings to coincide, necessitating renewal.
● Ignition timing incorrect (H100). Check the ignition timing as described in Chapter 3. Where no provision for adjustment exists, test the electronic ignition components and renew as required.
● Carburation fault. See Section 10.
● Mechanical resistance. Check that the brakes are not binding. On small machines in particular note that the increased rolling resistance caused by under-inflated tyres may impede acceleration.

Poor running or lack of power at high speeds

13 Weak spark at plug or erratic firing

● All items as for Section 9.
● HT lead insulation failure. Insulation failure of the HT lead and spark plug cap due to old age or damage can cause shorting when the engine is driven hard. This condition may be less noticeable, or not noticeable at all at lower engine speeds.

14 Fuel/air mixture incorrect

● All items as for Section 10, with the exception of items relative exclusively to low speed running.
● Main jet blocked. Debris from contaminated fuel, or from the fuel tank, and water in the fuel can block the main jet. Clean the fuel filter, the float bowl area, and if water is present, flush and refill the fuel tank.

● Main jet is the wrong size. The standard carburettor jetting is for sea level atmospheric pressure. For high altitudes, usually above 5000 ft, a smaller main jet will be required.
● Jet needle and needle jet worn. These can be renewed individually but should be renewed as a pair. Renewal of both items requires partial dismantling of the carburettor.
● Air bleed holes blocked. Dismantle carburettor and use compressed air to blow out all air passages.
● Reduced fuel flow. A reduction in the maximum fuel flow from the fuel tank to the carburettor will cause fuel starvation, proportionate to the engine speed. Check for blockages through debris or a kinked fuel line.

15 Compression low

● See Section 7.

Knocking or pinking

16 General causes

● Carbon build-up in combustion chamber. After high mileages have been covered large accumulations of carbon may occur. This may glow red hot and cause premature ignition of the fuel/air mixture, in advance of normal firing by the spark plug. Cylinder head removal will be required to allow inspection and cleaning.
● Fuel incorrect. A low grade fuel, or one of poor quality may result in compression induced detonation of the fuel resulting in knocking and pinking noises. Old fuel can cause similar problems. A too highly leaded fuel will reduce detonation but will accelerate deposit formation in the combustion chamber and may lead to early pre-ignition as described in item 1.
● Spark plug heat range incorrect. Uncontrolled pre-ignition can result from the use of a spark plug the heat range of which is too hot.
● Weak mixture. Overheating of the engine due to a weak mixture can result in pre-ignition occurring where it would not occur when engine temperature was within normal limits. Maladjustment, blocked jets or passages and air leaks can cause this condition.

Overheating

17 Firing incorrect

● Spark plug fouled, defective or maladjusted. See Section 5.
● Spark plug type incorrect. Refer to the Specifications and ensure that the correct plug type is fitted.
● Incorrect ignition timing. Timing that is far too much advanced or far too much retarded will cause overheating. Check the ignition timing is correct.

18 Fuel/air mixture incorrect

● Slow speed mixture strength incorrect. Adjust pilot air screw.
● Main jet wrong size. The carburettor is jetted for sea level atmospheric conditions. For high altitudes, usually above 5000 ft, a smaller main jet will be required.
● Air filter badly fitted or omitted. Check that the filter element is in place and that it and the air filter box cover are sealing correctly. Any leaks will cause a weak mixture.
● Induction air leaks. Check the security of the carburettor mountings and hose connections, and for cracks and splits in the hoses. Check also that the carburettor top is secure.
● Fuel level too low. See Section 3.
● Fuel tank filler cap air vent obstructed. Clear blockage.

19 Lubrication inadequate

● Oil pump settings incorrect. The oil pump settings are of great importance since the quantities of oil being injected are very small. Any variation in oil delivery will have a significant effect on the engine. Refer to Chapter 3 for further information.
● Oil tank empty or low. This will have disastrous consequences if left unnoticed. Check and replenish tank regularly.
● Transmission oil low or worn out. Check the level regularly and investigate any loss of oil. If the oil level drops with no sign of external leakage it is likely that the crankshaft main bearing oil seals are worn, allowing transmission oil to be drawn into the crankcase during induction.

20 Miscellaneous causes

● Engine fins clogged. A build-up of mud in the cylinder head and cylinder barrel cooling fins will decrease the cooling capabilities of the fins. Clean the fins as required.

Clutch operating problems

21 Clutch slip

● No clutch lever play. Adjust clutch cable play according to the procedure in Routine maintenance.
● Friction plates worn or warped. Overhaul clutch assembly, replacing plates out of specification.
● Steel plates worn or warped. Overhaul clutch assembly, replacing plates out of specification.
● Clutch spring broken or worn. Old or heat-damaged (from slipping clutch) springs should be replaced with new ones.
● Clutch inner cable snagging. Caused by a frayed cable or kinked outer cable. Replace the cable with a new one. Repair of a frayed cable is not advised.
● Clutch release mechanism defective. Worn or damaged parts in the clutch release mechanism could include the shaft, engine cover, pushrod or thrust bearing. Replace parts as necessary.
● Clutch centre and outer drum worn. Severe indentation by the clutch plate tangs of the channels in the centre and drum will cause snagging of the plates preventing correct engagement. If this damage occurs, renewal of the worn components is required.
● Lubricant incorrect. Use of a transmission lubricant other than that specified may allow the plates to slip.

22 Clutch drag

● Clutch lever play excessive. Adjust lever at bars or at cable end if necessary.
● Clutch plates warped or damaged. This will cause a drag on the clutch, causing the machine to creep. Overhaul clutch assembly.
● Clutch spring tension uneven. Usually caused by a sagged or broken spring. Check and replace springs.
● Transmission oil deteriorated. Badly contaminated transmission oil and a heavy deposit of oil sludge on the plates will cause plate sticking. The oil recommended for this machine is of the detergent type, therefore it is unlikely that this problem will arise unless regular oil changes are neglected.
● Transmission oil viscosity too high. Drag in the plates will result from the use of an oil with too high a viscosity. In very cold weather clutch drag may occur until the engine has reached operating temperature.
● Clutch centre and outer drum worn. Indentation by the clutch plate tangs of the channels in the centre and drum will

prevent easy plate disengagement. If the damage is light the affected areas may be dressed with a fine file. More pronounced damage will necessitate renewal of the components.
● Clutch outer drum seized to shaft. Lack of lubrication, severe wear or damage can cause the drum to seize to the shaft. Overhaul of the clutch, and perhaps the transmission, may be necessary to repair damage.
● Clutch release mechanism defective. Worn or damaged release mechanism parts can stick and fail to provide leverage. Overhaul clutch cover components.
● Loose or missing retaining circlip. Causes drum and centre misalignment, putting a drag on the engine. Engagement adjustment continually varies. Overhaul clutch assembly.

Gear selection problems

23 Gear lever does not return

● Weak or broken return spring. Renew the spring.
● Gearchange shaft bent or seized. Distortion of the gearchange shaft often occurs if the machine is dropped heavily on the gear lever. Provided that damage is not severe straightening of the shaft is permissible.

24 Gear selection difficult or impossible

● Clutch not disengaging fully. See Section 22.
● Gearchange shaft bent. This often occurs if the machine is dropped heavily on the gear lever. Straightening of the shaft is permissible if the damage is not too great.
● Gearchange claw arms or selector pins worn or damaged. Wear or breakage of any of these items may cause difficulty in selecting one or more gears. Overhaul the selector mechanism.
● Selector drum camplate or detent stopper arm damaged. Failure, rather than wear of these items may jam the drum thereby preventing gearchanging or causing false selection at high speed.
● Selector forks bent or seized. This can be caused by dropping the machine heavily on the gearchange lever or as a result of lack of lubrication. Though rare, bending of a shaft can result from a missed gearchange or false selection at high speed.
● Selector fork end and pin wear. Pronounced wear of these items and the grooves in the gearchange drum can lead to imprecise selection and, eventually, no selection. Renewal of the worn components will be required.
● Structural failure. Failure of any one component of the selector rod and change mechanism will result in improper or fouled gear selection.

25 Jumping out of gear

● Detent arm assembly worn or damaged. Wear of the arm roller and the cam with which it locates and breakage of the return spring can cause imprecise gear selection resulting in jumping out of gear. Renew the damaged components.
● Gear pinion dogs worn or damaged. Rounding off the dog edges and the mating recesses in adjacent pinion can lead to jumping out of gear when under load. The gears should be inspected and renewed. Attempting to reprofile the dogs is not recommended.
● Selector forks, drum and pinion grooves worn. Extreme wear of these interconnected items can occur after high mileages especially when lubrication has been neglected. The worn components must be renewed.
● Gear pinions, bushes and shafts worn. Renew the worn components.
● Bent gearchange shaft. Often caused by dropping the machine on the gear lever.

● Gear pinion tooth broken. Chipped teeth are unlikely to cause jumping out of gear once the gear has been selected fully; a tooth which is completely broken off, however, may cause problems in this respect and in any event will cause transmission noise.

26 Overselection

● Return spring weak or broken. Renew the spring.
● Detent assembly worn or broken. Renew the damaged items.
● Stopper arm spring worn or broken. Renew the spring.
● Selector claw arm stops worn. Repairs can be made by welding and reprofiling with a file.

Abnormal engine noise

27 Knocking or pinking

● See Section 16.

28 Piston slap or rattling from cylinder

● Cylinder bore/piston clearance excessive. Resulting from wear, or partial seizure. This condition can often be heard as a high, rapid tapping noise when the engine is under little or no load, particularly when power is just beginning to be applied. Reboring to the next correct oversize should be carried out, where possible, and a new oversize piston fitted; or the worn components should be renewed.
● Connecting rod bent. This can be caused by over-revving, trying to start a very badly flooded engine (resulting in a hydraulic lock in the cylinder) or by earlier mechanical failure. Attempts at straightening a bent connecting rod are not recommended. Careful inspection of the crankshaft should be made before renewing the damaged connecting rod.
● Gudgeon pin, piston boss bore or small-end bearing wear or seizure. Excess clearance or partial seizure between normal moving parts of these items can cause continuous or intermittent tapping noises. Rapid wear or seizure is caused by lubrication starvation.
● Piston rings worn, broken or sticking. Renew the rings after careful inspection of the piston and bore.

29 Other noises

● Big-end bearing wear. A pronounced knock from within the crankcase which worsens rapidly is indicative of big-end bearing failure as a result of extreme normal wear or lubrication failure. Remedial action in the form of a bottom end overhaul should be taken; continuing to run the engine will lead to further damage including the possibility of connecting rod breakage.
● Main bearing failure. Extreme normal wear or failure of the main bearings is characteristically accompanied by a rumble from the crankcase and vibration felt through the frame and footrests. Renew the worn bearings and carry out a very careful examination of the crankshaft.
● Crankshaft excessively out of true. A bent crank may result from over-revving or damage from an upper cylinder component or gearbox failure. Damage can also result from dropping the machine on either crankshaft end. Straightening of the crankshaft is not possible in normal circumstances; a replacement item should be fitted.
● Engine mounting loose. Tighten all the engine mounting nuts and bolts.
● Cylinder head gasket leaking. The noise most often as-

sociated with a leaking head gasket is a high pitched squeaking, although any other noise consistent with gas being forced out under pressure from a small orifice can also be emitted. Gasket leakage is often accompanied by oil seepage from around the mating joint or from the cylinder head holding down bolts and nuts. Leakage results from insufficient or uneven tightening of the cylinder head fasteners, or from random mechanical failure. Retightening to the correct torque figure will, at best, only provide a temporary cure. The gasket should be renewed at the earliest opportunity.
● Exhaust system leakage. Popping or crackling in the exhaust system, particularly when it occurs with the engine on the overrun, indicates a poor joint either at the cylinder port or at the exhaust pipe/silencer connection. Failure of the gasket or looseness of the clamp should be looked for.

Abnormal transmission noise

30 Clutch noise

● Clutch outer drum/friction plate tang clearance excessive.
● Clutch outer drum/spacer clearance excessive.
● Clutch outer drum/thrust washer clearance excessive.
● Primary drive gear teeth worn or damaged.
● Clutch shock absorber assembly worn or damaged.

31 Transmission noise

● Bearing or bushes worn or damaged. Renew the affected components.
● Gear pinions worn or chipped. Renew the gear pinions.
● Metal chips jammed in gear teeth.This can occur when pieces of metal from any failed component are picked up by a meshing pinion. The condition will lead to rapid bearing wear or early gear failure.
● Engine/transmission oil level too low. Top up immediately to prevent damage to gearbox and engine.
● Gearchange mechanism worn or damaged. Wear or failure of certain items in the selection and change components can induce mis-selection of gears (see Section 24) where incipient engagement of more than one gear set is promoted. Remedial action, by the overhaul of the gearbox, should be taken without delay.
● Chain snagging on cases or cycle parts. A badly worn chain or one that is excessively loose may snag or smack against adjacent components.

Exhaust smokes excessively

32 White/blue smoke (caused by oil burning)

● Piston rings worn or broken. Breakage or wear of any ring, but particularly the oil control ring, will allow engine oil past the piston into the combustion chamber. Examine and renew, where necessary, the cylinder barrel and piston.
● Cylinder cracked, worn or scored. These conditions may be caused by overheating, lack of lubrication, component failure or advanced normal wear. The cylinder barrel should be renewed and, if necessary, a new piston fitted.
● Oil pump settings incorrect. Check and reset the oil pump as described in Chapter 2.
● Crankshaft main bearing oil seals worn. Wear in the main bearing oil seals, often in conjunction with wear in the bearings themselves, can allow transmission oil to find its way into the crankcase and thence to the combustion chamber. This condition is often indicated by a mysterious drop in the transmission oil level with no sign of external leakage.

● Accumulated oil deposits in exhaust system. If the machine is used for short journeys only it is possible for the oil residue in the exhaust gases to condense in the relatively cool silencer. If the machine is then taken for a longer run in hot weather, the accumulated oil will burn off producing ominous smoke from the exhaust.

33 Black smoke (caused by over-rich mixture)

● Air filter element clogged. Clean or renew the element.
● Main jet loose or too large. Remove the float chamber to check for tightness of the jet. If the machine is used at high altitudes rejetting will be required to compensate for the lower atmospheric pressure.
● Cold start mechanism jammed on. Check that the mechanism works smoothly and correctly.
● Fuel level too high. The fuel level is controlled by the float height which can increase as a result of wear or damage. Remove the float bowl and check the float height. Check also that floats have not punctured; a punctured float will lose buoyancy and allow an increased fuel level.
● Float valve needle stuck open. Caused by dirt or a worn valve. Clean the float chamber or renew the needle and, if necessary, the valve seat.

Poor handling or roadholding

34 Directional instability

● Steering head bearing adjustment too tight. This will cause rolling or weaving at low speeds. Re-adjust the bearings.
● Steering head bearing worn or damaged. Correct adjustment of the bearing will prove impossible to achieve if wear or damage has occurred. Inconsistent handling will occur including rolling or weaving at low speed and poor directional control at indeterminate higher speeds. The steering head bearing should be dismantled for inspection and renewed if required. Lubrication should also be carried out.
● Bearing races pitted or dented. Impact damage caused, perhaps, by an accident or riding over a pot-hole can cause indentation of the bearing, usually in one position. This should be noted as notchiness when the handlebars are turned. Renew and lubricate the bearings.
● Steering stem bent. This will occur only if the machine is subjected to a high impact such as hitting a curb or a pot-hole. The bottom yoke/stem should be renewed; do not attempt to straighten the stem.
● Front or rear tyre pressures too low.
● Front or rear tyre worn. General instability, high speed wobbles and skipping over white lines indicates that tyre renewal may be required. Tyre induced problems, in some machine/tyre combinations, can occur even when the tyre in question is by no means fully worn.
● Swinging arm bearings worn. Difficulties in holding line, particularly when cornering or when changing power settings indicates wear in the swinging arm bearings. The swinging arm should be removed from the machine and the bearings renewed.
● Swinging arm flexing. The symptoms given in the preceding paragraph will also occur if the swinging arm fork flexes badly. This can be caused by structural weakness as a result of corrosion, fatigue or impact damage, or because the rear wheel spindle is slack.
● Wheel bearings worn. Renew the worn bearings.
● Loose wheel spokes. The spokes should be tightened evenly to maintain tension and trueness of the rim.
● Tyres unsuitable for machine. Not all available tyres will suit the characteristics of the frame and suspension, indeed, some tyres or tyre combinations may cause a transformation in the handling characteristics. If handling problems occur immediately after changing to a new tyre type or make, revert to the original tyres to see whether an improvement can be noted. In some instances a change to what are, in fact, suitable tyres may give rise to handling deficiences. In this case a thorough check should be made of all frame and suspension items which affect stability.

35 Steering bias to left or right

● Rear wheel out of alignment. Caused by uneven adjustment of chain tensioner adjusters allowing the wheel to be askew in the fork ends. A bent rear wheel spindle will also misalign the wheel in the swinging arm.
● Wheels out of alignment. This can be caused by impact damage to the frame, swinging arm, wheel spindles or front forks. Although occasionally a result of material failure or corrosion it is usually as a result of a crash.
● Front forks twisted in the steering yokes. A light impact, for instance with a pot-hole or low curb, can twist the fork legs in the steering yokes without causing structural damage to the fork legs or the yokes themselves. Re-alignment can be made by loosening the yoke pinch bolts, wheel spindle and mudguard bolts. Re-align the wheel with the handlebars and tighten the bolts working upwards from the wheel spindle. This action should be carried out only when there is no chance that structural damage has occurred.

36 Handlebar vibrates or oscillates

● Tyres worn or out of balance. Either condition, particularly in the front tyre, will promote shaking of the fork assembly and thus the handlebars. A sudden onset of shaking can result if a balance weight is displaced during use.
● Tyres badly positioned on the wheel rims. A moulded line on each wall of a tyre is provided to allow visual verification that the tyre is correctly positioned on the rim. A check can be made by rotating the tyre; any misalignment will be immediately obvious.
● Wheel rims warped or damaged. Inspect the wheels for runout as described in Chapter 5.
● Swinging arm bearings worn. Renew the bearings.
● Wheel bearings worn. Renew the bearings.
● Steering head bearings incorrectly adjusted. Vibration is more likely to result from bearings which are too loose rather than too tight. Re-adjust the bearings.
● Loose fork component fasteners. Loose nuts and bolts holding the fork legs, wheel spindle, mudguards or steering stem can promote shaking at the handlebars. Fasteners on running gear such as the forks and suspension should be check tightened occasionally to prevent dangerous looseness of components occurring.
● Engine mounting bolts loose. Tighten all fasteners.

37 Poor front fork performance

● Damping fluid level incorrect. If the fluid level is too low poor suspension control will occur resulting in a general impairment of roadholding and early loss of tyre adhesion when cornering and braking. Too much oil is unlikely to change the fork characteristics unless severe overfilling occurs when the fork action will become stiffer and oil seal failure may occur.
● Damping oil viscosity incorrect. The damping action of the fork is directly related to the viscosity of the damping oil. The lighter the oil used, the less will be the damping action imparted. For general use, use the recommended viscosity of oil, changing to a slightly higher or heavier oil only when a change in damping characteristic is required. Overworked oil, or

oil contaminated with water which has found its way past the seals, should be renewed to restore the correct damping performance and to prevent bottoming of the forks.
● Damping components worn or corroded. Advanced normal wear of the fork internals is unlikely to occur until a very high mileage has been covered. Continual use of the machine with damaged oil seals which allows the ingress of water, or neglect, will lead to rapid corrosion and wear. Dismantle the forks for inspection and overhaul.
● Weak fork springs. Progressive fatigue of the fork springs, resulting in a reduced spring free length, will occur after extensive use. This condition will promote excessive fork dive under braking, and in its advanced form will reduce the at-rest extended length of the forks and thus the fork geometry. Renewal of the springs as a pair is the only satisfactory course of action.
● Bent stanchions or corroded stanchions. Both conditions will prevent correct telescoping of the fork legs, and in an advanced state can cause sticking of the fork in one position. In a mild form corrosion will cause stiction of the fork thereby increasing the time the suspension takes to react to an uneven road surface. Bent fork stanchions should be attended to immediately because they indicate that impact damage has occurred, and there is a danger that the forks will fail with disastrous consequences.

38 Front fork judder when braking (see also Section 41)

● Wear between the fork stanchions and the fork legs. Renewal of the affected components is required.
● Slack steering head bearings. Re-adjust the bearings.
● Warped brake drum. If irregular braking action occurs fork judder can be induced in what are normally serviceable forks. Renew the damaged brake components.

39 Poor rear suspension performances

● Rear suspension unit damper worn out or leaking. The damping performance of most rear suspension units falls off with age. This is a gradual process, and thus may not be immediately obvious. Indications of poor damping include hopping of the rear end when cornering or braking, and a general loss of positive stability.
● Weak rear springs. If the suspension unit springs fatigue they will promote excessive pitching of the machine and reduce the ground clearance when cornering. Although replacement springs are available separately from the rear suspension damper unit it is probable that if spring fatigue has occurred the damper units will also require renewal.
● Swinging arm flexing or bearings worn. See Sections 34 and 36.
● Bent suspension unit damper rod. This is likely to occur only if the machine is dropped or if seizure of the piston occurs. If either happens the suspension units should be renewed as a pair.

Abnormal frame and suspension noise

40 Front end noise

● Oil level low or too thin. This can cause a 'spurting' sound and is usually accompanied by irregular fork action.
● Spring weak or broken. Makes a clicking or scraping sound. Fork oil will have a lot of metal particles in it.
● Steering head bearings loose or damaged. Clicks when braking. Check, adjust or replace.
● Fork clamps loose. Make sure all fork clamp pinch bolts are tight.

● Fork stanchion bent. Good possibility if machine has been dropped. Repair or replace tube.

41 Rear suspension noise

● Fluid level too low. Leakage of a suspension unit, usually evident by oil on the outer surfaces, can cause a spurting noise. The suspension units should be renewed as a pair.
● Defective rear suspension unit with internal damage. Renew the suspension units as a pair.

Brake problems

42 Brakes are spongy or ineffective

● Brake cable deterioration. Damage to the outer cable by stretching or being trapped will give a spongy feel to the brake lever. The cable should be renewed. A cable which has become corroded due to old age or neglect of lubrication will partially seize making operation very heavy. Lubrication at this stage may overcome the problem but the fitting of a new cable is recommended.
● Worn brake linings. Determine lining wear using the method described in Routine maintenance, or by removing the wheel and withdrawing the brake backplate. Renew the shoe/lining units as a pair if the linings are worn below the recommended limit.
● Worn brake camshaft. Wear between the camshaft and the bearing surface will reduce brake feel and reduce operating efficiency. Renewal of one or both items will be required to rectify the fault.
● Worn brake cam and shoe ends. Renew the worn components.
● Linings contaminated with dust or grease. Any accumulations of dust should be cleaned from the brake assembly and drum using a petrol dampened cloth. Do not blow or brush off the dust because it is asbestos based and thus harmful if inhaled. Light contamination from grease can be removed from the surface of the brake linings using a solvent; attempts at removing heavier contamination are less likely to be successful because some of the lubricant will have been absorbed by the lining material which will severely reduce the braking performance.

43 Brake drag

● Incorrect adjustment. Re-adjust the brake operating mechanism.
● Drum warped or oval. This can result from overheating or impact or uneven tension of the wheel spokes. The condition is difficult to correct, although if slight ovality only occurs, skimming the surface of the brake drum can provide a cure. This is work for a specialist engineer. Renewal of the complete wheel hub is normally the only satisfactory solution.
● Weak brake shoe return springs. This will prevent the brake lining/shoe units from pulling away from the drum surface once the brake is released. The springs should be renewed.
● Brake camshaft, lever pivot or cable poorly lubricated. Failure to attend to regular lubrication of these areas will increase operating resistance which, when compounded, may cause tardy operation and poor release movement.

44 Brake lever or pedal pulsates in operation

● Drums warped or oval. This can result from overheating or impact or uneven spoke tension. This condition is difficult to correct, although if slight ovality only occurs skimming the surface of the drum can provide a cure. This is work for a specialist engineer. Renewal of the hub is normally the only satisfactory solution.

45 Brake noise

● Drum warped or oval. This can cause intermittent rubbing of the brake linings against the drum. See the preceding Section.
● Brake linings glazed. This condition, usually accompanied by heavy lining dust contamination, often induces brake squeal. The surface of the linings may be roughened using glass-paper or a fine file.

46 Brake induced fork judder

● Worn front fork stanchions and legs, or worn or badly adjusted steering head bearings. These conditions, combined with uneven or pulsating braking as described in Section 44 will induce more or less judder when the brakes are applied, dependent on the degree of wear and poor brake operation. Attention should be given to both areas of malfunction. See the relevant Sections.

Electrical problems

47 Battery dead or weak

● Battery faulty. Battery life should not be expected to exceed 3 to 4 years, particularly where a starter motor is used regularly. Gradual sulphation of the plates and sediment deposits will reduce the battery performance. Plate and insulator damage can often occur as a result of vibration. Complete power failure, or intermittent failure, may be due to a broken battery terminal. Lack of electrolyte will prevent the battery maintaining charge.
● Battery leads making poor contact. Remove the battery leads and clean them and the terminals, removing all traces of corrosion and tarnish. Reconnect the leads and apply a coating of petroleum jelly to the terminals.
● Load excessive. If additional items such as spot lamps, are fitted, which increase the total electrical load above the maximum alternator output, the battery will fail to maintain full charge. Reduce the electrical load to suit the electrical capacity.
● Rectifier or resistor failure.
● Generator coils open-circuit or shorted.
● Charging circuit shorting or open circuit. This may be caused by frayed or broken wiring, dirty connectors or a faulty ignition switch. The system should be tested in a logical manner. See Section 50.

48 Battery overcharged

● Rectifier or resistor faulty. Overcharging is indicated if the battery becomes hot or it is noticed that the electrolyte level falls repeatedly between checks. In extreme cases the battery will boil causing corrosive gases and electrolyte to be emitted through the vent pipes.
● Battery wrongly matched to the electrical circuit. Ensure that the specified battery is fitted to the machine.

49 Total electrical failure

● Fuse blown. Check the main fuse. If a fault has occurred, it must be rectified before a new fuse is fitted.
● Battery faulty. See Section 47.
● Earth failure. Check that the frame main earth strap from the battery is securely affixed to the frame and is making a good contact.
● Ignition switch or power circuit failure. Check for current flow through the battery positive lead (red) to the ignition switch. Check the ignition switch for continuity.

50 Circuit failure

● Cable failure. Refer to the machine's wiring diagram and check the circuit for continuity. Open circuits are a result of loose or corroded connections, either at terminals or in-line connectors, or because of broken wires. Occasionally, the core of a wire will break without there being any apparent damage to the outer plastic cover.
● Switch failure. All switches may be checked for continuity in each switch position, after referring to the switch position boxes incorporated in the wiring diagram for the machine. Switch failure may be a result of mechanical breakage, corrosion or water.
● Fuse blown. Replace the fuse, if blown, only after the fault has been identified and rectified.

51 Bulbs blowing repeatedly

● Vibration failure. This is often an inherent fault related to the natural vibration characteristics of the engine and frame and is, thus, difficult to resolve. Modifications of the lamp mounting, to change the damping characteristics, may help.
● Intermittent earth. Repeated failure of one bulb, particularly where the bulb is fed directly from the generator, indicates that a poor earth exists somewhere in the circuit. Check that a good contact is available at each earthing point in the circuit.
● Reduced voltage. Do not overload the system with additional electrical equipment in excess of the system's power capacity and ensure that all circuit connections are maintained clean and tight.

HONDA H 100 & H 100 S SINGLES

Check list

Daily pre-ride checks

1 Check the level of engine oil in the tank
2 Make sure there is enough petrol in the tank to complete your journey
3 Check the operation of the front and rear brakes
4 Check the tyre pressures and remove any obstructions from the tread
5 Check the correct operation of the control cables and levers
6 Ensure the lights and speedometer function correctly

Weekly or every 150 miles (250 km)

1 Lubricate and adjust the final drive chain
2 Adjust the front and rear brakes
3 Examine the tyre treads for wear and damage and check the pressures
4 Check the correct operation of the front and rear suspension
5 Check the battery electrolyte level
6 Lubricate all control cables, levers and pivots

Monthly or every 600 miles (1000 km)

1 Check the gearbox oil level
2 Lubricate and adjust the final drive chain
3 Remove and clean the spark plug and reset the gap
4 Check and adjust points and ignition timing

Three monthly or every 1800 miles (3000 km)

1 Remove and clean the air filter element
2 Check the operation of the throttle and oil pump cables and adjust the oil pump
3 Inspect the condition of the wheels and check the wheel bearings for play

Six monthly or every 3600 miles (6000 km)

1 Adjust the carburettor
2 Change the transmission oil
3 Adjust the clutch cable
4 Renew the spark plug
5 Examine the condition of the brake shoes
6 Check the steering head bearings and the front fork bushes for wear and play
7 Decarbonize the engine and exhaust system

Annually or every 7200 miles (12 000 km)

1 Change the front fork oil
2 Check the fuel feed pipe for damage and leakage
3 Lubricate the instrument drive cables

Additional routine maintenance

1 Clean the fuel and oil filters
2 Grease the steering head bearings
3 Clean the machine

Adjustment data

Tyre pressures	Front	Rear
Solo	24 psi	28 psi
	(1.75 kg/cm²)	(2.00 kg/cm²)
Pillion	24 psi	32 psi
	(1.75 kg/cm²)	(2.25 kg/cm²)

Spark plug type — NGK BR7HS or ND W22FSR

Spark plug gap — 0.6 – 0.7 mm (0.024 – 0.028 in)

Ignition timing — 15° ± 3° BTDC (@ 3000 rpm – H100)

Idle speed — 1300 rpm

Contact breaker gap – H100 S — 0.35 mm (0.014 in)

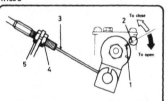

Oil pump cable adjustment

1 Operating lever reference mark
2 Pump body index mark
3 Operating cable
4 Locknut
5 Adjusting nut

Recommended lubricants

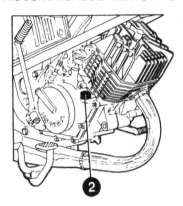

Component	Quantity	Grade
❶ Engine oil		
H100	1.6 lit (2.8 Imp pt)	Honda 2-stroke oil
H100 S	1.2 lit (2.1 Imp pt)	
❷ Transmission oil		
H100	0.9 lit (1.58 Imp pt)	SAE 10W/40 engine oil
H100 S	0.85 lit (1.50 Imp pt)	
❸ Front forks		
H100	113.5-118.5 cc (4.0-4.17 fl oz)	Fork oil or ATF
H100 S	80-86 cc (2.82-3.03 fl oz)	
❹ Final drive chain	As required	Aerosol chain lubricant
❺ Wheel bearings	As required	High melting point grease
❻ Steering head bearings	As required	High melting point grease
❼ Pivot points	As required	High melting point grease
❽ Control cable	As required	Light machine oil

ROUTINE MAINTENANCE GUIDE

For specifications and information relating to the H100 S II model, refer to Chapter 7

Routine maintenance

For specifications and information relating to the H100 S II model, refer to Chapter 7

Periodic routine maintenance is a continuous process which should commence immediately the machine is used. The object is to maintain all adjustments and to diagnose and rectify minor defects before they develop into more extensive, and often more expensive, problems.

It follows that if the machine is maintained properly, it will both run and perform with optimum efficiency, and be less prone to unexpected breakdowns. Regular inspection of the machine will show up any parts which are wearing, and with a little experience, it is possible to obtain the maximum life from any one component, renewing it when it becomes so worn that it is liable to fail.

Regular cleaning can be considered as important as mechanical maintenance. This will ensure that all the cycle parts are inspected regularly and are kept free from accumulations of road dirt and grime.

Cleaning is especially important during the winter months, despite its appearance of being a thankless task which very soon seems pointless. On the contrary, it is during these months that the paintwork, chromium plating, and the alloy casings suffer the ravages of abrasive grit, rain and road salt. A couple of hours spent weekly on cleaning the machine will maintain its appearance and value, and highlight small points, like chipped paint, before they become a serious problem.

The various maintenance tasks are described under their respective mileage and calendar headings, and are accompanied by diagrams and photographs where pertinent.

It should be noted that the intervals between each maintenance task serve only as a guide. As the machine gets older, or if it is used under particularly arduous conditions, it is advisable to reduce the period between each check.

For ease of reference, most service operations are described in detail under the relevant heading. However, if further general information is required, this can be found under the pertinent Section heading and Chapter in the main text.

Although no special tools are required for routine maintenance, a good selection of general workshop tools is essential. Included in the tools must be a range of metric ring or combination spanners, a selection of crosshead screwdrivers,

and two pairs of circlip pliers, one external opening and the other internal opening. Additionally, owing to the extreme tightness of most casing screws on Japanese machines, an impact screwdriver, together with a choice of large or small cross-head screw bits, is absolutely indispensable. This is particularly so if the engine has not been dismantled since leaving the factory.

Daily (pre-riding check)

Before taking the machine out on the road there are certain checks which should be completed to ensure that it is in a safe and legal condition to be used.

1 Engine oil level

On H100 models there is an oil level gauge on the top of the petrol/oil tank whch provides an instant indication of the amount of oil remaining in the tank. The oil level should never be allowed to drop into the red zone of the gauge. If it is, check the oil tank/oil pump feed pipe for air bubbles, and bleed the system if necessary. This procedure is described in full in Section 19 of Chapter 2. On H100 S models the oil tank is mounted behind the right-hand side panel, the low level line marked on the tank being visible via the inspection aperture in the side panel. Do not allow the oil level to fall below this line. To top up the tank, release its retaining catch and withdraw the side panel, then unscrew the tank retaining knob to allow the tank to swing out on its hinge. Remove the filler cap.

The engine oil level should be checked at the beginning of each journey, and at each refuelling stop if the journey is a long one. Never allow the oil level to run low as the first warning you will get if the tank runs dry is when the engine seizes up through lack of lubrication.

Use only two-stroke oil designed for motorcycle injection systems when topping-up the tank, and only fill the tank to the bottom of the filler neck.

Do not allow needle to drop into red zone of the oil level gauge
– H100

Use only two-stroke oil when topping up

Do not allow level to fall below minimum oil level line – H100 S

Rotate screw to release when removing sidepanel – H100 S

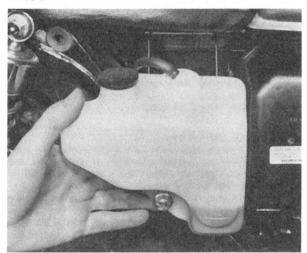

Unscrew chromed knob to swing out oil tank for topping up – H100 S

Check tyre pressures with an accurate gauge

2 Petrol level

Checking the petrol level may seem obvious, but it is all too easy to forget. Ensure that you have enough petrol to complete your journey, or at least to get you to the nearest petrol station.

3 Brakes

Check that the front and rear brakes work effectively and without binding. Ensure that the cable and the rod linkage are correctly adjusted and that they are properly lubricated. Ensure that both stop lamp switches are functioning correctly.

4 Tyres

Check the tyre pressures with a gauge that is known to be accurate. It is worthwhile purchasing a pocket gauge for this purpose because the gauges on garage forecourt airlines are notoriously inaccurate. The pressures should always be checked with the tyres cold. Even a few miles travelled will warm up the tyres to a point where pressures increase and an inaccurate reading will result. Tyre pressures are:

	Front	Rear
Solo	24 psi (1.75 kg/cm²)	28 psi (2.00 kg/cm²)
Pillion	24 psi (1.75 kg/cm²)	32 psi (2.25 kg/cm²)

At the same time, examine the tyres themselves. Look for damage, especially splitting of the sidewalls. Remove any small stones or other road debris caught between the treads. Any such debris will work its way into the tyre and may penetrate the inner tube, thus causing rapid deflation and a resultant loss of control. The tendency to pick up road debris is much more pronounced in the rear tyre and this, therefore, should be checked particularly carefully. The depth of tread remaining should also be measured in view of both the legal and safety aspects. It is vital to keep the tread depth above the UK legal limit of 1 mm of depth over three-quarters of the tread breadth around the entire circumference of the tyre. Many riders, however, consider nearer 2 mm to be the limit for secure roadholding, traction, and braking, especially in adverse weather conditions.

5 Controls

Check the throttle and clutch cables and levers, the gear lever and the footrests to ensure that they are adjusted correctly, functioning correctly, and that they are securely fastened. If a bolt is going to work loose, or a cable snap, it is better that it is discovered at this stage, with the machine at a standstill, rather than when it is being ridden.

6 Lights and speedometer

Check that all lights, flashing indicators, horn and speedometer are working correctly to make sure that the machine complies with all legal requirements in this respect.

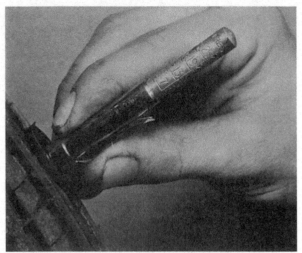

Keep a careful check on the amount of tyre tread remaining

Weekly, or approximately every 150 miles (250 km)

It is with the weekly inspection of the machine that the proper procedure of routine maintenance starts. The daily checks serve to ensure merely that the machine is in a safe and legal condition, and contribute little to maintenance other than to give the owner an accurate picture of what items need attention. In this way the daily checks, if done conscientiously, will give early warning of any faults which are about to appear. For this reason, carry out the daily checks before performing the weekly maintenance tasks listed below.

1   Lubricating and adjusting the final drive chain
Due to the fully enclosed chaincase fitted to the H100A, the need for constant chain maintenance is greatly reduced as the chain is not exposed to rain and road dirt. This, however, does not mean that the chain can be forgotten. Regular applications of lubricant are still required to keep it in good condition. Remove the chaincase inspection cap and examine the chain, spinning the back wheel to ensure that the whole length of the chain is seen. If the rollers appear in any way dry, the chain must be lubricated, and if there are signs of extreme lack of lubrication, such as rusty areas or kinks in the chain, the chain should be removed from the machine for thorough cleaning and lubrication, using a special chain grease such as Linklyfe or Chainguard. This procedure, while long and potentially messy, is essential if the chain is to be correctly lubricated and is to last for as long as possible.

Lubrication, by removing the chain and immersing it in molten Linklyfe or Chainguard, should be regarded as a regular part of routine maintenance and should be carried out at intervals of 500 - 1000 miles, this interval depending on the degree and severity of use to which the machine is put. The full procedure is described in detail in Section 11 of Chapter 5.

For weekly chain lubrication, however, a more satisfactory method is the application of grease by one of the proprietary brands of aerosol lubricant. This method is much cleaner and quicker than that given above, but should only be considered an addition to the use of Linklyfe or Chainguard, and not an alternative to it. Lubricant applied by aerosol can will not be able to penetrate the inner bushes and bearing surfaces of the chain as effectively as molten grease. The use of engine oil must be considered as a last resort if nothing else is available as it is easily flung off the chain and is too thin to be of any real value in lubricating the chain bearing surfaces.

The chain adjustment procedure is covered in full under the monthly maintenance heading. The actual interval depends very much on the use to which the machine is put and must be determined by the owner.

An aerosol provides the most convenient form of chain lubrication – apply at least every week

Note particularly that on H100 S models, the lack of a full chaincase will mean that the chain must be lubricated and adjusted at more frequent intervals. The procedure is precisely as that given above for the H100, but note the following differences: Check the chain tension midway between the sprockets on the chain lower run. The wheel spindle nut is not fitted with a securing split-pin, and since there is no separate sprocket carrier, there is no larger retaining sleeve nut.

2   Adjusting the brakes
Good brakes are essential to the safety of both machine and rider and should be checked constantly to ensure that they remain in peak condition. Adjustment is extremely simple for both the front and the rear brakes.

The front brake is adjusted by means of the adjusting nut on the extreme lower end of the front brake cable. Turn the nut clockwise to tighten up the brake until there is 20 – 30 mm ($\frac{3}{4}$ – 1$\frac{1}{4}$ in) free play measured at the tip of the brake lever, before the brake begins to operate.

The rear brake is adjusted by means of the adjusting nut fitted to the rear end of the brake operating rod. Turn the nut clockwise to reduce free play until there is 20 – 30 mm ($\frac{3}{4}$ – 1$\frac{1}{4}$ in) measured at the brake pedal tip before the brake begins to operate. Remember that the stop lamp switch setting will be altered by adjusting the rear brake. To adjust the switch setting, turn the plastic sleeve nut as necessary until the stop lamp bulb lights just as the brake pedal has taken up its free play and is beginning to engage the shoes on the brake drum.

Complete the weekly brake maintenance by oiling all lever pivot points, all exposed lengths of cable, cable nipples and the rear brake linkage. Do not allow excessive amounts of oil to get on to the operating arms, in case any should find its way into the brake drum or on to the tyre.

3   Checking the tyres
Although the daily check will give an accurate indication of the general condition of the tyres, a more thorough check is recommended each week. Check that the tyre pressures are correct, as given in the daily test, and carefully examine the treads and sidewalls, as previously mentioned. Be careful to remove any foreign matter from the treads.

4   Checking the front and rear suspension
Ensure that the front forks operate smoothly and progressively by pumping them up and down whilst the front brake is held on. Any faults revealed by this check should be investigated further, because any deterioration in the handling of the machine can have serious consequences if left un-

Front brake is adjusted at the nut on the lower end of the cable

Rear brake is adjusted at the nut on the end of the brake rod

Adjust stop lamp switch whenever the rear brake adjustment is altered

Electrolyte level should be maintained between level marks on battery casing

remedied. Check the condition of the fork stanchions. As with most current production machines, the fork stanchions are left exposed in the interests of fashion, and are thus prone to damage from stone chips or abrasion. Any damage to the stanchions will lead to rapid wear of the fork seals and can only be cured by renewing the stanchions. This is both costly and time consuming, so it is worth checking that the area below each dust seal is kept clean and greased. Remove any abrasive grit which may have accumulated around the dust seal lip. The above problems can be eliminated by fitting fork gaiters, these being available from most accessory stockists.

The rear suspension can be checked with the machine on the centre stand. Check that all of the suspension components are securely attached to the frame. Check for free play in the swinging arm by pushing and pulling it horizontally.

5 Checking the battery

Remove the right-hand (left-hand, H100 S) side panel and check that the electrolyte levels are between the level marks on the battery casing. Top up with distilled water to the upper level if necessary. Check that the connections are clean and tight and that the battery breather pipe is free from kinks and blockages.

6 General checks and lubrication

Check around the machine, looking for loose nuts, bolts or screws, retightening them as necessary. Check the stand and

lever pivots for security and lubricate them with light machine oil or engine oil. Make sure that the stand springs are in good condition.

It is advisable to lubricate the handlebar switches and stop lamp switches with WD40 or a similar water dispersant lubricant. This will keep the switches working properly and prolong their life, especially if the machine is used in adverse weather conditions.

Monthly or every 600 miles (1000 km)

First complete the tasks listed under the daily and weekly checks, and then carry out the following:

1 Checking the gearbox oil level

Before checking the gearbox oil level, the engine must be started and warmed up to normal operating temperature. Place the machine on its centre stand so that it is upright on level ground. Unscrew the level plug which is situated in the right-hand engine cover, immediately in from of the kickstart shaft. Oil should slowly trickle out if the level is correct. Top up through the filler plug if necessary, but be careful not to overfill. If oil flows out in a steady stream, allow it to drain until the flow slows to a gentle trickle. Use a good quality SAE 10W/40 engine oil when topping up, and ensure that the gearbox filler, level, and drain plugs are securely tightened before finishing.

Keep handlebar switches lubricated with silicone-based aerosol spray

Oil should trickle slowly out if oil level is correct

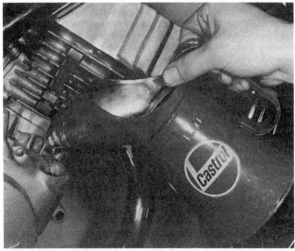

Use only recommended grade of oil when topping up

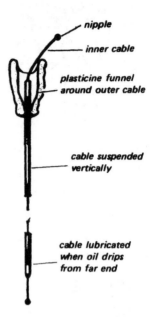

- nipple
- inner cable
- plasticine funnel around outer cable
- cable suspended vertically
- cable lubricated when oil drips from far end

Oiling a control cable

2 Adjusting and lubricating the final drive chain

Place the machine on its centre stand on level ground with the transmission in neutral. Chain adjustment must be carried out at regular intervals to compensate for wear. Due to the fact that this wear never takes place evenly along the length of the chain, tight spots will develop which must be allowed for during adjustment. Remove the chaincase inspection cap. Rotate the rear wheel slowly, testing the chain tension at points along the entire length of the chain, until the tightest spot is found. There should be 10 – 20 mm ($\frac{3}{8}$ – $\frac{3}{4}$ inch) total up and down movement, or free play, in the chain at this spot.

To adjust the chain, withdraw the split pin from the wheel spindle nut, and slacken both the spindle nut and the larger sprocket retaining sleeve nut by just enough to permit the wheel to be moved backwards. Tighten the nut on each of the chain adjusters to pull the wheel spindle backwards, thus tightening the chain until the correct amount of free play is achieved.

When adjusting the chain tension, it is essential to preserve correct rear wheel alignment by pulling back each side of the wheel spindle by exactly the same amount. Vertical lines stamped in the swinging arm serve as reference marks to aid in this. The index mark stamped in each chain adjuster must be aligned with the same reference mark on each side of the swinging arm.

Once the chain is correctly adjusted, tighten the sprocket retaining sleeve nut securely and, using a torque wrench, tighten the rear wheel spindle nut to 5.5 – 6.5 kgf m (40 – 47 lbf ft). Check rear brake adjustment and ensure that all nuts and bolts are securely fastened. Use a new split pin to retain the spindle nut, spreading the ends of the split pin to secure it. Ensure that the rear wheel rotates freely and easily. Complete chain lubrication, if necessary, as previously described.

3 Cleaning and resetting the spark plug

Detach the spark plug cap, and using the correct spanner remove the spark plug. Clean the electrodes using a wire brush followed by a strip of fine emery cloth or paper. Check the plug gap with a feeler gauge, adjusting it if necessary to within the range of 0.6 – 0.7 mm (0.024 – 0.028 in). Make adjustments by bending the outer electrode, never the inner (central) electrode.

Before fitting the spark plug smear the threads with a graphited grease; this will aid subsequent removal.

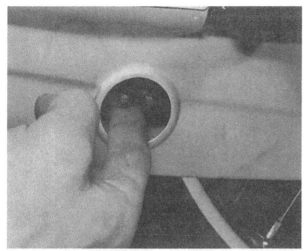

Find tightest spot in the chain ...

... slacken rear wheel spindle nut ...

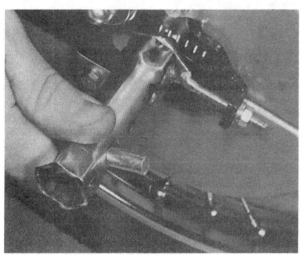

... and the larger sprocket retaining nut ...

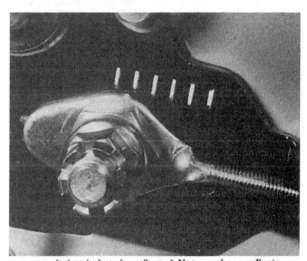

... to permit the chain to be adjusted. Note marks on adjuster and swinging arm which aid wheel alignment

The spark plug should be regularly cleaned and the gap checked

Three monthly or every 1800 miles (3000 km)

First complete all the operations listed under the previous mileage/time headings and then carry out the following:

1 Cleaning the air filter element

It is vitally important that the air filter element is kept clean and in good condition if the engine is to function properly. If the element becomes choked with dust it follows that the airflow to the engine will be impaired, leading to poor performance and high fuel consumption. Conversely, a damaged filter will allow excessive amounts of unfiltered air to enter the engine, which can result in an increased rate of wear and possibly damage due to the weak nature of the mixture. The interval specified above indicates the maximum time limit between each cleaning operation. Where the machine is used in particularly adverse conditions it is advised that cleaning takes place on a much more frequent basis.

On H100 models, remove the left-hand side panel/air filter cover which is secured by three screws, then remove the single retaining screw and withdraw the metal supporting frame and the element itself. On H100 S models, release its retaining catch and withdraw the left-hand side panel, then withdraw the air filter cover which is retained by three screws. Withdraw the element and its supporting frame.

The foam can be cleaned by washing in a high flash point solvent, such as white spirit. The use of petrol (gasoline) is not approved by the manufacturer in view of the potential fire risk. Allow the element to dry, then impregnate the foam with SAE 80 or 90 gear oil, removing any excess by squeezing it out. The element can now be reassembled and fitted.

If inspection has revealed any holes or tears, the element must be renewed immediately. On no account be tempted to omit the element in view of the damage that may ensue from the resulting weak mixture.

2 Checking the adjustment and condition of the throttle and oil pump cables

The throttle and oil pump cables must be regularly inspected to ensure that the rider can control the engine's speed accurately and that the correct amount of lubricant reaches the engine via the oil pump. Turn the handlebars from one full lock position to the other to ensure that the cable does not foul and that engine speed at tickover does not increase. Check the outer cables for signs of damage, then examine the exposed portions of the inner cables. Any signs of kinking or fraying will indicate that renewal is required. To obtain maximum life and reliability from the cables they should be thoroughly lubricated. To do the job properly and quickly use one of the hydraulic cable oilers available from most motorcycle shops. Free one end of the cable and assemble the cable oiler as described by the manufacturer's instructions. Operate the oiler until oil emerges from the lower end, indicating that the cable is lubricated throughout its length. This process will expel any dirt or moisture and will prevent its subsequent ingress.

If a cable oiler is not available, an alternative is to remove the cable from the machine. Hang the cable upright and make up a small funnel arrangement using plasticene or by taping a plastic bag around the upper end. Fill the funnel with oil and leave it overnight to drain through. Note that where nylon-lined cables are fitted, they should be used dry or lubricated with a silicone-based lubricant suitable for this application. On no account use ordinary engine oil because this will cause the liner to swell, pinching the cable.

Check all pivots and control levers, cleaning and lubricating them to prevent wear or corrosion. Where necessary, dismantle and clean any moving part which may have become stiff in operation. If cable removal is necessary for lubrication purposes, read Chapter 2 very carefully first, with special reference to Sections 10 and 18 regarding correct adjustment of the carburettor and the oil pump.

Cable adjustment must be made first at the throttle cable, and then at the oil pump cable. Open and close the throttle several times, allowing it to snap shut under its own pressure. Ensure that it is able to shut off quickly and fully at all handlebar positions. Check that there is 2 – 6 mm (0.08 – 0.24 in) free play measured around the circumference of the inner flange of the rubber twistgrip. If not, use the adjuster on the carburettor top to achieve the correct setting, completing the operation, if necessary, with the adjuster at the handlebar twistgrip. Open and close the throttle again to settle the cable and to check that adjustment is not disturbed. Once the throttle cable is correctly adjusted, open the throttle fully and check that the reference mark etched on the oil pump control lever lines up exactly with the fixed index mark on the pump body. If adjustment is necessary, slacken the adjuster locknut and turn the adjusting nut until the marks line up exactly. Remember that the throttle must be fully open to align the marks. Once the oil pump cable is correctly adjusted, tighten the adjuster locknut and open the throttle fully two or three times to check the lever operation and to ensure that the adjustment remains constant.

Complete the routine oil pump maintenance by carefully examining the oil lines. Check that the fitting clamps and wire clips are securely fastened and ensure that the piping is not chafing on any part of the frame or engine. Any damaged piping must be immediately renewed. Look carefully for leaks, air bubbles or dirt in the oil lines. Refer to the relevant Sections of

Chapter 2 for details of filter cleaning, oil system bleeding and any other work which may be necessary.

The air filter element must be regularly cleaned and re-oiled

Throttle cable free play is measured around circumference of twistgrip

Oil pump setting marks (arrowed) must align exactly with throttle in fully open position

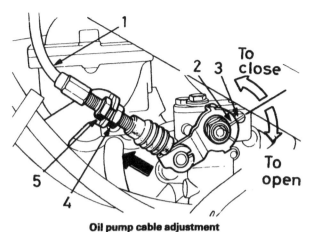

Oil pump cable adjustment

1	*Operating cable*	4	*Locknut*
2	*Control lever reference mark*	5	*Adjusting nut*
3	*Index mark*		

3 Checking the condition of the wheels and wheel bearings

Place the machine on the centre stand so that the front wheel is raised clear of the ground. Spin the wheel and check the rim alignment. Small irregularities can be corrected by tightening the spokes in the affected area although a certain amount of experience is necessary to prevent over-correction. Any flats in the wheel rim will be evident at the same time. These are more difficult to remove and in most cases it will be necessary to have the wheel rebuilt on a new rim. Apart from the effect on stability, a flat will expose the tyre bead and walls to greater risk of damage if the machine is run with a deformed wheel.

Check for loose and broken spokes. Tapping the spokes is the best guide to tension. A loose spoke will produce a quite different sound and should be tightened by turning the nipple in an anti-clockwise direction. Always check for run-out by spinning the wheel again. If the spokes have to be tightened by an excessive amount, it is advisable to remove the tyre and tube as detailed in Chapter 5. This will enable the protruding ends of the spokes to be ground off, thus preventing them from chafing the inner tube and causing punctures. The condition of the rear wheel can be checked in exactly the same way as described above.

To check the condition of the wheel bearings, grasp the wheel at its rim and try to move it to and fro at right angles to the normal direction of rotation. If any free play at all is felt, the wheel concerned must be removed, dismantled, and its bearings examined and renewed if necessary as described in Sections 4 or 8 of Chapter 5.

4 Checking and adjusting the contact breaker points and ignition timing – H100 S

Remove the four retaining screws and withdraw the left-hand engine cover, then disconnect and remove the spark plug. Examine the faces of the contact breaker points via the aperture in the generator rotor, when the points are fully open. If they are smooth and level but grey in colour, use fine emery or similar to polish the faces and restore good contact; light pitting can be removed by the careful use of a fine file. If the contact faces are heavily burnt or pitted, if they have worn at an uneven angle or if the heel of the moving point is so badly worn that correct ignition timing cannot be maintained, the points must be renewed.

Remove the generator rotor as described in Section 12 of Chapter 1, then slacken the small screw and nut which clamps the low tension lead terminal to the contact breaker unit. The terminal is forked and may be pulled from position without

removing the screw completely. Note the position of the terminal in relation to the insulating washers on the screw; this will aid reassembly. Remove the contact breaker adjustment screw to free the contact breaker assembly from the stator plate. Fit the new contact breaker assembly, then refit the rotor as described in Section 28 of Chapter 1, and adjust the points gap.

If the contacts are in good condition, measure the gap using a feeler gauge. A 0.35 mm (0.014 in) gauge should be a light sliding fit – the points **must be** within the range: 0.3 - 0.4 mm (0.012 - 0.016 in). Should they require adjustment, slacken the securing screw *just* enough to permit the fixed contact to be moved, using a small screwdriver. Tighten the securing screw and then recheck the gap. Because no provision is given for adjustment of the ignition timing the point at which firing occurs is dependent on the contact breaker gap. Because of this, if contact breaker adjustment is made the ignition timing must be checked as a matter of course to determine whether it is still accurate.

To check the ignition timing, use a battery and bulb test circuit or a multimeter set to the most sensitive resistance scale. Disconnect the contact breaker low tension lead (black/yellow wire) at the connector joining it to the main wiring loom. The meter or test circuit negative (-) terminal is connected to a good earth point on the frame or engine, as shown in the accompanying illustration. When connected in this fashion, the bulb will dim (or go out) or the meter needle will flicker, to indicate increased resistance as the contact breaker points open; this should be when the 'F' mark stamped in the rotor rim aligns exactly with the raised index mark on the crankcase wall in approximately the 2 o'clock position.

Rotate the generator rotor anti-clockwise until the marks align and check that the points open at exactly that moment. If the ignition timing is incorrect slacken the contact breaker gap adjustment screw and adjust the contact breaker so that the bulb goes out at the correct point; opening the gap advances the ignition timing, closing the gap retards it. Retighten the screw. Rotate the rotor until the points are fully open and then check the gap. If the gap is within the specified range of 0.3 - 0.4 mm (0.012 - 0.016 in) all is well. If, however, the gap is outside the range it is evident that the contact breakers have worn to a point where correct ignition timing and correct contact breaker clearance cannot be maintained simultaneously. If this is the case the contact breaker assembly must be renewed regardless of its apparent condition.

If the equipment is available, the ignition timing can be checked dynamically, ie with the engine running, as outlined in Section 9 of Chapter 3. Using a strobe lamp as described, check that the timing marks align at 1500 rpm. If not, stop the engine and make the necessary adjustments as described above.

Complete contact breaker maintenance by applying a few drops of light oil to the felt lubricating wick, but do not apply too much or the surplus may foul the contact breaker faces.

Use screwdriver as shown to adjust contact breaker gap

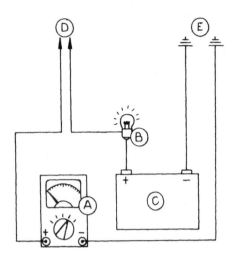

Checking the static ignition timing – H100 S

A   Multimeter
B   Bulb
C   Battery
D   Black/yellow wire terminal
E   Frame or engine earth

Six monthly or every 3600 miles (6000 km)

First complete all the operations listed under the previous mileage/time headings and then carry out the following.

1 Carburettor adjustment

If rough running of the engine has developed some adjustment of the carburettor pilot setting and tick-over speed may be required. If this is the case refer to Chapter 2, Section 10 for details. Do not make these adjustments unless they are obviously required, there is little to be gained by unwarranted attention to the carburettor. Complete carburettor maintenance by slackening the drain screw on the float chamber, turning the petrol on, and allowing a small amount of fuel to drain through, thus flushing any water or dirt from the carburettor. Tighten the drain screw securely and switch the petrol off.

2 Changing the transmission oil

Routine changes of the transmission oil are made much more effective if the oil is hot. At high temperatures the oil is thinner and flows more easily; thus draining faster and more completely and taking all the minute particles of metal and dirt with it. It is evident from this fact that the engine/gearbox unit must be at normal working temperature when the transmission oil is drained. To achieve this, drain the oil the night before the maintenance work is to be carried out, immediately after the machine has been ridden. This ideal situation will allow the oil to drain completely and the engine to cool down, thus preventing the risk of personal injury. Place the machine on its centre stand on level ground, remove filler, level, and drain plugs and allow the oil to drain into a suitably-sized container. When the oil is completely drained, check that the drain plug sealing washer is in good condition and replace the drain plug, tightening it to a torque setting of 2.0 – 2.5 kgf m (14 – 18 lbf ft). Refill the gearbox with 0.9 lit (1.58 pint) on H100 models or 0.85 lit (1.50 pint) on H100 S models, of good quality SAE 10W/40 engine oil, then refit the filler plug. Start the engine and allow it to warm up by running it for a few minutes. Stop the engine and recheck the oil level, adding oil if necessary. Ensure filler, level, and drain plugs are securely tightened.

Ensure that drain plug washer is in good condition before replacing drain plug

3 Clutch adjustment

Accurate adjustment of the clutch cable is necessary to ensure efficient operation of the unit. There is no provision for adjustment of the clutch itself.

Normal adjustment is made using the cable lower end adjuster, at the operating arm on the crankcase cover. Slacken the locknut and turn the adjuster nut. This tensions the inner cable and reduces the free play at the handlebar lever. When the free play at the handlebar lever is between 10 – 20 mm (0.4 – 0.8 in), the adjustment is correct. Retighten the lower end locknut.

4 Renewing the spark plug

The manufacturer recommends that the spark plug is renewed as a precautionary measure at this stage. Always ensure that a plug of the correct type and heat range is fitted, and that the gap is set to the prescribed 0.6 – 0.7 mm (0.024 – 0.028 in) prior to installation. If the old plug is in reasonable condition, it can be cleaned and re-gapped and carried as an emergency spare in the toolbox.

5 Examining the condition of the brake shoes

Close periodic inspection of the brake shoes is necessary to ensure continued braking efficiency. The regular checks every week will ensure that the brakes are kept in proper adjustment but it is advisable to remove the wheels to check the amount of friction material left. It should be noted that while no external wear limit indicators are provided, there is a simple check which will give a rough guide to the amount of friction material remaining. If the angle formed between the operating arm on the brake backplate and the brake cable or brake rod is more than 90° when the brake is fully applied, the shoes are worn out and must be renewed immediately.

This, however, only reveals the thickness of friction material, not its condition. It is therefore necessary to remove the wheels and withdraw the brake components for cleaning and examination. Refer to Chapter 5, Sections 3 and 7 for instructions on wheel removal, and Sections 5 and 9 of the same Chapter for details of brake examination. Pay special attention to cleaning any foreign material from the shoes if they are sufficiently unworn to be re-usable and ensure that the brake cams are properly greased.

6 Checking the steering head bearings and front fork bushes

Wear or play in the steering head bearings will cause imprecise handling and can be dangerous if allowed to develop unchecked. Test for play by pushing and pulling on the handlebars whilst holding the front brake on. Any wear in the head races will be seen as movement between the fork yokes and the steering lug, but be careful not to mistake play in the fork bushes as steering head bearing wear. Check, if in doubt, by gripping the front wheel between the knees and rocking the handlebars from side to side. If any play is felt in the fork legs, the forks must be dismantled for the bushes and bearing surfaces to be examined and renewed if necessary, as described in the relevant Sections of Chapter 4. If the steering head bearings are confirmed as worn, the free play can be removed by adjustment.

Before carrying out adjustment, place a wooden crate or similar item beneath the crankcase so that the front wheel is raised clear of the ground. Check that the handlebars will turn smoothly and freely from lock to lock. If the steering feels notchy or jerky in operation it may be due to worn or damaged bearings. Should this be suspected it will be necessary to overhaul the steering head bearings as described in Chapter 4.

To adjust the steering head bearings, slacken the large steering stem nut at the centre of the top fork yoke, then use a C-spanner to tighten the slotted adjuster nut immediately below the top yoke. As a guide to adjustment, tighten the slotted nut until a light resistance is felt, then back it off by $\frac{1}{8}$ turn. The object is to remove all discernible play without applying any appreciable preload. It should be noted that it is possible to apply a loading of several tons on the small steering head bearings without this being obvious when turning the handlebars. This will cause an accelerated rate of wear, and thus must be avoided.

7 Decarbonising the engine and exhaust system

Due to the lubrication system employed on two-stroke engines there is a surplus of oil present in the combustion chamber which is not completely burned during combustion. This surplus oil manifests itself as a relatively rapid build-up of carbon in the combustion chamber, exhaust port, and exhaust system. Regular decarbonising of the engine and exhaust is therefore necessary. The exact interval depends very much on the way in which the machine is used, as an engine which is used on short journeys only, or run at consistently low speeds, will build up carbon deposits at a faster rate, and therefore need more frequent decarbonising, than an engine run at higher speeds or used for longer journeys. It is therefore recommended that the owner carries out a full decarbonising operation every six months, or at the distance specified, until he has acquired enough experience to decide for himself when decarbonising is actually necessary.

To carry out the operation, slacken and remove the two exhaust front mounting nuts and the rear mounting bolts. Withdraw the exhaust system. Slacken and remove the two carburettor mounting nuts and very gently slide the carburettor back off its mounting studs. Slacken and remove the engine top mounting bolt and remove the spark plug cap and spark plug. Progressively slacken the four cylinder head sleeve nuts to ensure an even release of pressure, remove the nuts and withdraw the cylinder head and its gasket. Remove the oil feed pipe from the inlet stub and plug it immediately using a screw or bolt of suitable size to prevent the loss of oil and the entry of any dirt or air. Turn the crankshaft using the kickstarter until the piston reaches the top of its stroke. Remove the cylinder barrel, noting that a few gentle taps with a soft-faced mallet around the area of the cylinder base may be necessary to free the joint. Before pulling the barrel clear of the piston, carefully pack the crankcase mouth with clean rag to prevent the entry of any dirt.

Using a blunt-edged scraping tool to prevent scratching the soft alloy components, carefully remove all traces of carbon from the cylinder head combustion chamber, the piston crown, the exhaust port and from the front length of the exhaust pipe. Finish off using a soft rag and metal polish to give a smooth, polished finish to these areas which will reduce the ability of future carbon deposits to adhere so easily. Check that the piston rings are free in their grooves in the piston. If necessary, very carefully remove the rings using three thin strips of metal to free them if they are stuck in place. Clean the skirt of the piston and remove any carbon in the ring grooves using a section of broken piston ring to ensure that the grooves are not damaged. Carefully examine the rings as described in Section 18 of Chapter 1 and renew them if necessary. Should removal of the piston be considered necessary, be sure to obtain new circlips to replace those disturbed.

Carefully clean the gasket surfaces of the crankcase, cylinder barrel, and cylinder head. Ensure that the piston and rings, if disturbed, have been correctly refitted, and oil them liberally. Place a new cylinder base gasket in position and replace the cylinder barrel, taking great care that the piston ring end gaps are correctly located at their respective pegs, and that the piston rings are not broken as the barrel is replaced. Once the piston and rings are securely in the cylinder bore, remove the rag from the crankcase mouth and slide the barrel into place. Place a new cylinder head gasket in position and replace the cylinder head. Tighten the four cylinder head nuts progressively and evenly to a torque setting of 1.9 – 2.3 kgf m (14 – 17 lbf ft). Replace the carburettor on its studs and securely tighten the retaining nuts. Unplug the oil feed line and replace it on the inlet stub union. If it has been properly plugged, bleeding of the oil injection system will not be necessary. In such a case, oil will be visible in the end of the pipe. If not, carry

Steering head bearing adjustment is made by turning the slotted adjuster nut ...

... with a C-spanner

out the second part of the bleeding operation as described in Section 19 of Chapter 2. Replace the spark plug and cap, and the engine top mounting bolt. Tighten the engine mounting bolt to a torque setting of 3.0 – 4.0 kgf m (22 – 29 lbf ft). Using a new exhaust gasket, replace the exhaust system. Tighten the mounting nuts and bolt by hand only at first, then use a spanner to securely tighten first the front mounting nuts, and then the single rear mounting bolt; this procedure being necessary to ensure that the exhaust system is correctly aligned on its mountings and under no strain when those mountings are tightened down.

The exhaust system is fitted with a removable baffle tube secured by a single screw at the extreme rear of the silencer. Slacken and remove the screw and withdraw the baffle tube. If the carbon deposits are of an oily nature only, it will usually suffice to scrub the baffle with a wire brush and a petrol/paraffin mixture. If, however, the carbon deposits are hard and dry, the use of a blow lamp will be necessary to burn the carbon away. Whichever method is used, the holes in the baffle tube must be completely free and the external surfaces clean and polished before the tube is refitted. Note that if the tube proves difficult to remove, it will most probably have been stuck firmly in place by carbon deposits. Removal in such cases is very difficult and some method should be improvised of removing the baffle while damaging either it or the silencer as little as possible. In such a case, baffle removal must be carried out at more frequent intervals in order to keep the deposits to a minimum.

Once the engine and exhaust components have been dismantled, cleaned, and correctly rebuilt, the engine can be started again. Remember to carry out the oil pump bleeding procedure if necessary and note that even if no new components have been fitted during decarbonising, the parts cleaned will have been disturbed from their positions and will need time to settle down again.

Annually or every 7200 miles (12 000 km)

Once a year the machine must be taken off the road for a major maintenance session which will involve first carrying out all the previous maintenance tasks listed under the five mileage/time headings. To these operations must be added the following:

1 Changing the fork oil

This is an important task which must be carried out to ensure the continuing stability and safety of the machine on the road. Front fork oil, like any other oil, gradually degenerates as it loses viscosity and becomes contaminated with water and dirt; this degeneration causing a very gradual loss of damping which may not be noticed by the rider until the machine is actually unsafe to ride. Regular changes of the fork oil will eliminate this possibility and provide a convenient moment to inspect closely the front suspension components.

Unfortunately no drain plugs are fitted to the front forks on this machine and so complete removal of each fork leg is necessary. Each fork leg can then be inverted and the fork oil allowed to drain. Pump the leg to complete the draining process, noting that this task is made easier by first removing the spring retaining plug and the fork spring. Details of the full fork leg removal procedure are given in Sections 2 and 3 of Chapter 4 with details of examination and repair work being given in Sections 7 and 8. Reassembly and refitting are covered in Sections 9 and 10 respectively. While it is by no means necessary to dismantle the fork legs in order to change the fork oil, if there is any doubt about the condition of the fork bearing surfaces or the oil seals, the necessary work would most conveniently be carried out at this stage.

When draining and inspection are complete, reassemble the fork legs using the specified amount of oil. The oil recommended is ATF, but there are many proprietary brands of oil

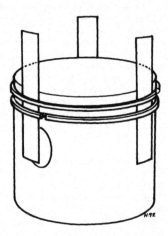

Method of removing and replacing piston rings

developed especially for use in forks which are available as alternatives.

2 Checking the fuel pipe condition

Give the pipe which connects the fuel tap and carburettor a close visual examination, checking for cracks or any signs of leakage. In time, the synthetic rubber pipe will tend to deteriorate, and will eventually leak. Apart from the obvious fire risk, the evaporating fuel will affect fuel economy. If the pipe is to be renewed, always use the correct replacement type to ensure a good leak-proof fit. Never use natural rubber tubing because this will tend to break up when in contact with petrol, and will obstruct the carburettor jets.

3 Lubricating the instrument cables

The speedometer cable is retained at both ends by a knurled, threaded ring. Using a suitable pair of pliers, slacken both rings and withdraw the cable. On H100 S models, it will be necessary to remove the headlamp unit from the headlamp casing to gain access to the cable upper end. Remove the inner cable by pulling it out from the bottom of the outer. Carefully examine the inner cable for signs of fraying, kinking, or for any shiny areas which will indicate tight spots, and the outer cable for signs of cracking, kinking or any other damage. Renew either cable if necessary. To lubricate the cable, smear a small quantity of grease on to the lower length only of the inner. Do not allow any grease on the top six inches of the cable as the grease will work its way up the length of the cable as it rotates and get into the speedometer head. This will rapidly ruin the instrument, necessitating its renewal. When lubrication is complete, insert the inner back into the outer cable and replace it on the machine.

On H100 S models, remove fully the clamp bolt at its lower end to release the tachometer cable, then remove it for checking and greasing as described above.

Additional routine maintenance

Certain aspects of routine maintenance make it impossible to place operations under specific mileage or calendar headings, or may necessitate modification of these headings. A good example of the latter is the effect of a dusty environment on certain maintenance operations, notably cleaning the air filter element. Similarly a machine ridden over rough or dirty roads will require more frequent attention to the cycle parts, ie suspension components and wheels, and to the chain. A machine ridden at constant high speeds will need attention to the brakes and engine/transmission components far more than a machine ridden at excessively slow speeds. The latter will, however, need more frequent decarbonising. The problem of how to achieve the correct balance between too little maintenance, which will result in premature and expensive damage to the machine, and too much, is a delicate one which is unfortunately only resolved by personal experience. This experience is best gained by strict adherence to the specified mileage/time headings until the owner feels qualified to alter them to suit his own machine.

Some components will require inspection and attention at intervals dependent on usage rather than mileage or age. Two such tasks are given below:

1 Cleaning the fuel and oil filters

It should be necessary to attend to the filters only on rare occasions, for example when traces of dirt or water are found continually in the carburettor, or when dirt is seen in the oil tank or even in the oil tank/oil pump feed line. Carefully read Chapter 2, with special reference to Sections 2, 3, 5 and 15 which are concerned with removal of the petrol/oil tank and cleaning of the tank and filters.

2 Greasing the steering head bearings

As already mentioned under the six monthly or 3600 mile (6000 km) heading, the steering head bearings must be checked, and adjusted if necessary, at regular intervals. If, however, they have not been dismantled during the course of routine maintenance or for accident repairs, they should be dismantled, examined, and packed with new grease every two years. This operation is quite lengthy, involving removal of the front forks, and would therefore fit in most conveniently with the annual fork oil change. Carefully read Chapter 4 with special reference to Sections 2, 3, 4 and 5 which describe the full procedure in detail.

3 Cleaning the machine

Keeping the motorcycle clean should be considered as an important part of the routine maintenance, to be carried out whenever the need arises. A machine cleaned regularly will not only succumb less speedily to the inevitable corrosion of external surfaces, and hence maintain its market value, but will be far more approachable when the time comes for maintenance or service work. Furthermore, loose or failing components are more readily spotted when not partially obscured by a mantle of road grime and oil.

Surface dirt should be removed using a sponge and warm, soapy water; the latter being applied copiously to remove the particles of grit which might otherwise cause damage to the paintwork and polished surfaces.

Oil and grease is removed most easily by the application of a cleaning solvent such as 'Gunk' or 'Jizer'. The solvent should be applied when the parts are still dry and worked in with a stiff brush. Large quantities of water should be used when rinsing off, taking care that water does not enter the carburettors, air cleaners or electrics.

If desired a polish such as Solvol Autosol can be applied to the aluminium alloy parts to restore the original lustre. This does not apply in instances, much favoured by Japanese manufacturers, where the components are lacquered. Application of a wax polish to the cycle parts and a good chrome cleaner to the chrome parts will also give a good finish. Always wipe the machine down if used in the wet, and make sure the chain is well oiled. There is less chance of water getting into control cables if they are regularly lubricated, which will prevent stiffness of action.

Standard torque settings

Specific torque settings will be found at the end of the specifications section of each chapter. Where no figure is given, bolts should be secured according to the table below.

Fastener type (thread diameter)	kgf m	lbf ft
5 mm bolt or nut	0.45 – 0.6	3.5 – 4.5
6 mm bolt or nut	0.8 – 1.2	6 – 9
8 mm bolt or nut	1.8 – 2.5	13 – 18
10 mm bolt or nut	3.0 – 4.0	22 – 29
12 mm bolt or nut	5.0 – 6.0	36 – 43
5 mm screw	0.35 – 0.5	2.5 – 3.6
6 mm screw	0.7 – 1.1	5 – 8
6 mm flange bolt	1.0 – 1.4	7 – 10
8 mm flange bolt	2.4 – 3.0	17 – 22
10 mm flange bolt	3.0 – 4.0	22 – 29

Chapter 1 Engine, clutch and gearbox

For specifications and information relating to the H100 S II model, refer to Chapter 7

Contents

Specifications

Engine

Type	Air-cooled, single cylinder, two-stroke
Bore	50.5 mm (1.988 in)
Stroke	49.5 mm (1.949 in)
Capacity	99 cc (6.04 cu in)

Compression ratio:
    H100 .... 6.7 : 1
    H100 S .... 7.2 : 1
Lubrication system .... Honda 2-stroke oil injection
Port timing:

	H100	H100 S
Intake	Opening and closing reed valve controlled	
Exhaust opens at	90.85° BBDC	85.00° BBDC
Exhaust closes at	90.85° ABDC	85.00° ABDC
Scavenge opens at	57.50° BBDC	58.00° BBDC
Scavenge closes at	57.50° ABDC	58.00° ABDC
Booster opens at	59.50° BBDC	61.00° BBDC
Booster closes at	59.50° ABDC	61.00° ABDC

Cylinder head

Maximum warpage .... 0.10 mm (0.004 in)

Piston

OD .... 50.455 – 50.470 mm (1.9864 – 1.9870 in)
Wear limit .... 50.420 mm (1.9850 in)
Gudgeon pin bore ID .... 14.002 – 14.008 mm (0.5513 – 0.5515 in)
Wear limit .... 14.030 mm (0.5524 in)
Gudgeon pin OD .... 13.994 – 14.00 mm (0.5509 – 0.5512 in)
Wear limit .... 13.980 mm (0.5504 in)
Piston/gudgeon pin clearance .... 0.002 – 0.014 mm (0.0001 – 0.0006 in)
Wear limit .... 0.040 mm (0.0016 in)

Piston rings

End gap (installed) .... 0.1 – 0.25 mm (0.004 – 0.010 in)
Wear limit .... 0.55 mm (0.022 in)

Cylinder barrel

Bore .... 50.50 – 50.52 mm (1.9882 – 1.9890 in)
Wear limit .... 50.57 mm (1.9909 in)
Piston/cylinder clearance .... 0.030 – 0.060 mm (0.0012 – 0.0024 in)
Wear limit .... 0.100 mm (0.0039 in)
Compression pressure .... 13 kg/cm² (185 psi)

Crankshaft

Connecting rod small-end ID .... 19.005 – 19.017 mm (0.7482 – 0.7487 in)
Wear limit .... 19.030 mm (0.7492 in)
Connecting rod big-end side clearance .... 0.15 – 0.55 mm (0.006 – 0.022 in)
Wear limit .... 0.85 mm (0.033 in)
Connecting rod big-end radial play (max) .... 0.05 mm (0.002 in)
Run-out at journals (max) .... 0.10 mm (0.004 in)

Primary drive

Type .... Gear
Reduction ratio .... 4.117 : 1 (70/17)

Clutch

Type .... Wet, multiplate
Number of plates:
    Plain 3
    Friction 4
Friction plate thickness .... 2.9 – 3.0 mm (0.114 – 0.118 in)
Wear limit .... 2.5 mm (0.098 in)
Plain plate warpage (max) .... 0.2 mm (0.008 in)
Clutch spring free length .... 28.1 mm (1.11 in)
Wear limit .... 27.1 mm (1.07 in)
Clutch outer bush ID .... 17.000 – 17.018 mm (0.6693 – 0.6700 in)
Wear limit .... 17.060 mm (0.6717 in)

Kickstart mechanism

Shaft OD .... 11.966 – 11.984 (0.4711 – 0.4718 in)
Wear limit .... 11.950 mm (0.4704 in)
Pinion ID .... 12.016 – 12.034 mm (0.4731 – 0.4738 in)
Wear limit .... 12.070 mm (0.4752 in)
Idler gear ID .... 15.032 – 15.050 mm (0.5918 – 0.5925 in)

Wear limit .... 15.100 mm (0.5945 in)
Output shaft OD .... 15.000 – 15.018 mm (0.5906 – 0.5913 in)
Wear limit .... 14.940 mm (0.5882 in)

Crankshaft balancer
Balancer idler gear shaft OD .... 9.972 – 9.987 mm (0.3926 – 0.3932 in)
Wear limit .... 9.930 mm (0.3909 in)

Oil pump
Oil pump driven gear shaft OD .... 9.965 – 9.987 mm (0.3923 – 0.3932 in)
Wear limit .... 9.930 mm (0.3909 in)

Gearbox
Type .... 5 speed constant mesh
Reduction ratios:
    1st 3.083 (37/12)
    2nd 1.882 (32/17)
    3rd 1.400 (28/20)
    4th 1.130 (26/23)
    5th 0.960 (24/25)
Selector fork ID .... 10.000 – 10.018 mm (0.3937 – 0.3944 in)
Wear limit .... 10.05 mm (0.3957 in)
Selector fork shaft OD .... 9.972 – 9.987 mm (0.3926 – 0.3932 in)
Wear limit .... 9.950 mm (0.3917 in)
Selector fork claw thickness .... 4.93 – 5.00 mm (0.194 – 0.197 in)
Wear limit .... 4.50 mm (0.1772 in)
Input shaft OD .... 16.966 – 16.984 mm (0.6680 – 0.6687 in)
Wear limit .... 16.930 mm (0.6665 in)
Output shaft OD:
    At 16.5 mm shaft diameter .... 16.466 – 16.484 mm (0.6483 – 0.6490 in)
    Wear limit .... 16.440 mm (0.6472 in)
    At 17 mm shaft diameter .... 16.978 – 16.989 mm (0.6684 – 0.6688 in)
    Wear limit .... 16.960 mm (0.6677 in)
    At 19 mm shaft diameter .... 18.959 – 18.980 mm (0.7464 – 0.7472 in)
    Wear limit .... 18.930 mm (0.7453 in)
Output shaft 2nd gear bush:
    OD .... 19.984 – 19.995 mm (0.7868 – 0.7872 in)
    Wear limit .... 19.975 mm (0.7864 in)
    ID .... 17.016 – 17.034 mm (0.6699 – 0.6706 in)
    Wear limit .... 17.050 mm (0.6713 in)
Gear pinion inside diameters:
    Input shaft 4th and 5th .... 17.016 – 17.034 mm (0.6699 – 0.6706 in)
    Wear limit .... 17.100 mm (0.6732 in)
    Output shaft 1st .... 16.522 – 16.543 mm (0.6505 – 0.6513 in)
    Wear limit .... 16.600 mm (0.6535 in)
    Output shaft 2nd .... 20.020 – 20.041 mm (0.7882 – 0.7890 in)
    Wear limit .... 20.060 mm (0.7898 in)
    Output shaft 3rd .... 19.020 – 19.041 mm (0.7488 – 0.7496 in)
    Wear limit .... 19.100 mm (0.7520 in)
Selector drum OD:
    At 13 mm drum diameter .... 12.934 – 12.984 mm (0.5092 – 0.5112 in)
    Wear limit .... 12.850 mm (0.5059 in)
    At 36 mm drum diameter .... 35.950 – 35.975 mm (1.4154 – 1.4163 in)
    Wear limit .... 35.900 mm (1.4134 in)

Final drive
Type .... Chain and sprockets
Reduction ratio:
    H100 .... 2.333 : 1 (35/15)
    H100 S .... 2.266 : 1 (34/15)
Chain size .... 428 ($\frac{1}{2}$ x $\frac{5}{16}$ in) x 108 links

Torque wrench settings
Component	kgf m	lbf ft
Cylinder head nut	1.9 – 2.3	14 – 17
Generator rotor nut	6.0 – 7.0	43 – 50
Primary drive pinion nut	4.5 – 5.5	33 – 40
Engine mounting bolt	3.0 – 4.0	22 – 29
Oil drain plug	2.0 – 2.5	14 – 18
Kickstart and gearchange lever pinch bolt	0.8 – 1.2	6 – 9

1 General description

These models employ an air-cooled, single cylinder, two-stroke engine built in unit with the gearbox and clutch assemblies. The engine castings are of light alloy construction and the cylinder barrel incorporates a steel liner.

The crankshaft is of conventional construction, incorporating caged needle roller bearings at the small-end and big-end with journal ball main bearings. Primary drive is direct from the crankshaft pinion to the clutch outer drum. To counter the imbalance inherent in all single cylinder engines a gear driven single shaft primary balancer is used. This is driven through a balancer idler gear which incorporates an anti-backlash gear to minimise noise.

Engine power is transmitted through the clutch to a five-speed constant mesh gearbox which is lubricated by splash from its own oil reservoir contained within the crankcase castings. The engine is lubricated by Honda's own two-stroke injection system in which a pump, driven from the clutch outer drum, feeds a metered amount of oil straight into the induction tract, thus eliminating the need for using a petrol/oil mixture. The oil pump is interconnected to the throttle cable, oil delivery thus being controlled by both throttle position and engine speed.

2 Operations with the engine/gearbox unit in the frame

The following items can be overhauled with the engine/gearbox unit installed in the frame:

a) Cylinder head
b) Cylinder barrel, piston, and reed valve assembly
c) Clutch and primary drive
d) Kickstart mechanism
e) Gear selector mechanism
f) Oil pump and drive components
g) Final drive sprocket
h) Flywheel generator components

When several operations need to be undertaken simultaneously, it would probably be an advantage to remove the complete unit from the frame; a very simple operation which should take approximately half an hour. This will give the advantage of better access and more working space.

3 Operations with the engine/gearbox unit removed from the frame

It will be necessary to remove the engine/gearbox unit from the frame and separate the crankcase halves to gain access to the following:
a) Crankshaft assembly
b) Balancer shaft
c) Gearbox components

4 Removing the engine/gearbox unit from the frame

1 Before commencing any dismantling work, it will be necessary to drain the gearbox oil. As oil is thinner when hot and therefore flows better than when cold, it is best to drain the oil when the engine is at normal operating temperature. It is recommended that the oil is drained immediately after the machine has completed a journey, and preferably the night before work is due to commence, to achieve the most complete draining possible. This procedure has the added advantage of ensuring that there is no risk of personal injury due to burns inflicted by hot engine components, as these will have been able to cool down overnight. To drain the oil, place the machine on its centre stand, slacken and remove the gearbox filler and level plugs, which are situated in the right-hand outer engine cover, and place a container underneath the engine which must be large enough to catch the oil. The gearbox contains approximately 1 litre (1.76 Imp pint) of oil which is released by slackening and removing the hexagon-headed drain plug situated centrally on the underside of the crankcase.

2 Place the machine securely on its centre stand ensuring that it is in no danger of falling while work is in progress. To prevent the machine from rolling forward, place wooden blocks securely against the front wheel. Working on the machine is made considerably easier if the machine is raised a few feet from the ground on a stout wooden bench or similar support.

3 On H100 models, remove the right-hand side panel by removing its mounting screw and disengaging it from its rear mounting. On H100 S models, rotate the catch to release it and lift away the left-hand side panel. Disconnect the two battery leads at their respective snap connectors, unclip the rubber battery retaining strap and withdraw the battery and vent tube. The battery should be replaced safely to one side to await reassembly, but remember that if it is anticipated that the machine is to be out of service for an extended period of time, arrangements must be made to give the battery a refresher charge every month or so as detailed in Chapter 6. Trace the main lead up from the generator to the wiring loom, disconnect the wires at their snap connectors and release any cable clips which may secure the main lead to the frame. Remove the spark plug cap and secure it and the HT lead clear of the engine.

4 Slacken and remove the pinch bolt securing the kickstart lever to its shaft, and remove the kickstart lever. Marking the lever and shaft prior to removal will aid correct position on reassembly. Remove the clutch cable by slackening the adjuster locknuts and sliding the cable clear of the adjuster bracket. The cable end nipple can now be disengaged from the operating arm on the right-hand outer engine cover. Tape the cable to the frame so that it is clear of the engine. On H100 S models, remove fully its clamp bolt and pull the tachometer drive cable out of its housing on the right-hand outer casing.

5 On H100 S models, remove the single bolt securing the silencer front mounting to the footrest bracket. On all models, remove the two nuts retaining the exhaust pipe to the cylinder barrel, then take the weight of the exhaust system and remove the single bolt securing the system to the frame. Carefully remove the exhaust system from the machine. Replace the two nuts on their studs and the bolt in its position in the frame to prevent their loss.

6 Slacken and remove the two nuts securing the carburettor flange to the inlet stub. Push the carburettor back along the mounting studs as far as possible. When the engine is removed the carburettor will remain in place in the frame with the air filter and cable connections undisturbed.

7 Moving round to the left-hand side of the machine, disconnect the oil pump cable by slackening the locknut and adjusting nut and sliding the cable clear of the adjuster bracket. The cable end nipple can now be disengaged from the operating arm on the pump. Secure the cable clear of the engine. Disconnect the oil tank/oil pump feed line at the junction just below the petrol/oil tank, placing a finger over the pipe end temporarily to stop the flow of oil. Both pipes can then be plugged using screws or bolts of suitable size to prevent the loss of oil from the tank and the entry of dirt or air into the oil tank/oil pump feed line. Care must be taken not to damage the pipe ends.

8 Slacken and remove the pinch bolt which retains the gearchange lever and remove the lever. Marking the lever and shaft prior to removal will aid correct positioning on reassembly. Remove the footrest bar by slackening and removing the three

securing bolts. Note that all three are removed from the left, one screwing into the frame horizontally from the outside next to the sidestand pivot, one vertically on the left-hand side, and the last being concealed underneath the engine crankcase and screwing into the frame horizontally between the footrest bar and centre stand pivot.

9   Unclip the black plastic chain cover from its position at the rear of the engine unit, slacken and remove the four bolts securing the left-hand engine cover and remove the cover. Slacken and remove the two gearbox sprocket retaining bolts locking the sprocket by applying the rear brake if necessary. Turn the sprocket retaining plate until it can be pulled off the splines. Remove the gearbox sprocket by sliding it clear of the shaft complete with the chain. Disengage the sprocket from the chain and remove it, leaving the chain to hang around the swinging arm pivot. Should there not be enough slack in the drive chain to allow the sideways displacement of the sprocket, the chain must first be disconnected at its connecting link.

10   Return to the right-hand side of the machine and slacken, then remove the nuts which fasten the three engine mounting bolts. The engine/gearbox unit is now ready to be removed, but it is possible that more clearance may be required to prevent the rear brake pedal fouling the engine. If this is necessary first disconnect the stop lamp switch by unhooking its spring from the pedal, and then slacken off the rear brake adjuster as required.

11   To remove the engine/gearbox unit one person only is needed, but the assistance of a second person would be helpful. Push out first the engine bottom mounting bolt and then the upper rear mounting bolt, noting that the latter passes through two separate spacers, H100 only, one on each side of the engine unit. These spacers will drop clear as the bolt is withdrawn, and the engine unit swings forward, and must be replaced on the bolt immediately to minimise confusion on reassembly. As the upper rear bolt is withdrawn, hold the engine/gearbox unit with one hand and allow it to swing slowly forwards, while guiding the carburettor off its mounting studs with the other. While this can be done by one person taking great care, it is at this point that assistance from a second person would be helpful. With one person removing the engine mounting bolts while the other supports the engine and guides the carburettor as described, the whole operation becomes much easier and the risk of losing or damaging any components is greatly lessened. Whichever method is used, the engine/gearbox unit should now be hanging free in the frame supported only by its top mounting bolt. Support the weight of the unit in one hand, withdraw the bolt with the other and remove the unit from the frame. Replace the engine mounting bolts in position in the frame to prevent their loss.

4.1 Drain plug is situated centrally in the underside of the crankcase

4.3 Disconnect the generator leads at their snap connectors

4.4 Remove clamp bolt fully to release tachometer cable – H100 S

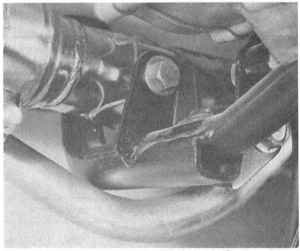

4.5 H100 S silencer is retained by single bolt at front

4.7 Disconnect oil feed line at union just beneath the tank

4.8 Footrest mounting bolts – H100 S

4.9 Remove retaining screws to release left-hand engine cover

5 Dismantling the engine/gearbox unit: preliminaries

1 Before any dismantling work is undertaken, the external surfaces of the unit should be thoroughly cleaned and degreased. This will prevent the contamination of the engine internals, and will also make working a lot easier and cleaner. A high flash point solvent, such as paraffin (kerosene) can be used, or better still, a proprietary engine degreaser such as Gunk. Use old paintbrushes and toothbrushes to work the solvent into the various recesses of the engine castings. Take care to exclude solvent or water from the electrical components and inlet and exhaust ports. The use of petrol (gasoline) as a cleaning medium should be avoided, because the vapour is explosive and can be toxic if used in a confined space.
2 When clean and dry, arrange the unit on the workbench leaving a suitable clear area for working. Gather a selection of small containers and plastic bags so that parts can be grouped together in an easily identifiable manner. Some paper and a pen should be on hand to permit notes to be made and labels attached where necessary. A supply of clean rag is also required.
3 Before commencing work, read through the appropriate section so that some idea of the necessary procedure can be gained. When removing the various engine components it should be noted that great force is seldom required, unless

specified. In many cases, a component's reluctance to be removed is indicative of an incorrect approach or removal method. If in any doubt, re-check with the text.

6 Dismantling the engine/gearbox unit: removing the cylinder head, barrel, and piston

1 As mentioned earlier in this Chapter, removal of the cylinder head, barrel, and piston, is possible with the engine/gearbox unit in or out of the frame but removal of the engine top mounting bolt, the exhaust and carburettor will first be necessary.
2 Release the four cylinder head sleeve nuts. These should be slackened progressively, by about one turn at a time, in a diagonal sequence. This will avoid any possibility of the cylinder head casting warping.
3 Remove the cylinder head and gasket. If, for any reason the joint will not free easily, do not use force to remove the cylinder head and do not risk damaging the gasket surface or the castings themselves by attempting to lever the joint apart. Instead tap round the head joint with a soft-faced mallet to jar it free. Care must be taken not to damage the cooling fins which are brittle and easy to break.
4 Remove the oil pump/inlet stub feed pipe from its union on the inlet stub and plug it using a screw or bolt of suitable size. This will prevent the loss of oil and the entry of dirt or air into the injection system. Turn the crankshaft until the piston reaches the top of its stroke. Gently ease the barrel along its studs, freeing the joint if necessary by tapping around the cylinder base with a soft-faced mallet. Before pulling the barrel clear of the piston, carefully pack the crankcase mouth with clean rag to catch any debris or broken piston rings which might otherwise fall into the crankcase. Remove the cylinder barrel.
5 The reed valve/inlet stub assembly is removed by slackening the four retaining bolts in a diagonal sequence. Again tap the joint surface areas with a soft-faced mallet to jar them free and be especially careful not to damage the castings, which are very light, and therefore delicate in construction. Put the reed valve assembly to one side and ensure that it is not dropped or allowed to get dirty.
6 Prise out one of the gudgeon pin circlips using a small electrical screwdriver or similar tool in the slot provided. Support the piston and push the gudgeon pin out until the connecting rod is freed. The piston can now be lifted away. If the gudgeon pin is a tight fit, warm the piston crown with a rag soaked in boiling water, and tap the gudgeon pin out using a hammer and a soft metal drift of suitable size. Be very careful to support the piston and connecting rod during this operation

and do not use excessive force, or there is a risk of bending the connecting rod. Remove the small-end bearing.

7   Place all the components to one side for further attention, but discard any circlips disturbed during dismantling. These should never be re-used, new ones must be obtained and fitted on reassembly.

7   Dismantling the engine/gearbox unit: removing the right-hand outer casing

1   The right-hand outer casing can be removed with the engine/gearbox unit in or out of the frame. In the former instance it will be necessary to carry out the following operations prior to removal:

a) Drain the gearbox oil
b) Remove the kickstart lever
c) Disconnect the clutch cable from its operating arm
d) Remove the exhaust system – H100 S
e) Remove the tachometer drive cable – H100 S

The above operations are all described in Section 4 of this Chapter.

6.2 Slacken cylinder head nuts progressively and evenly

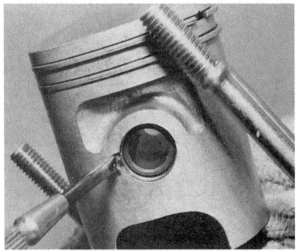

6.6 Prise out gudgeon pin circlip using a suitable tool

2   Slacken and remove the ten hexagon-headed screws from the periphery of the casing. As they are of various lengths, the best method of keeping the bolts is to mark a piece of thick cardboard with the positions of the bolts and then to push each bolt through the cardboard in its correct relative position as it is removed. This ensures that each bolt is kept in the same position that it occupies in the crankcase and greatly simplifies the task of correctly replacing them. Be careful to place the clutch cable adjuster bracket with its respective bolts on the cardboard.

3   When all the bolts have been removed, lift the casing away, taking care not to damage the kickstart shaft oil seal as it passes over the shaft splines. Be prepared to catch any residual oil left in the casing. There is a thrust washer on the kickstart shaft which may stick to the outer cover or fall clear. This should be replaced on the kickstart shaft for safekeeping.

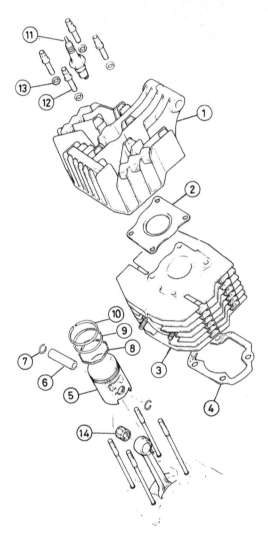

Fig. 1.1 Cylinder head and barrel

1	Cylinder head	9	Second ring
2	Cylinder head gasket	10	Top ring
3	Cylinder barrel	11	Spark plug
4	Cylinder base gasket	12	Sleeve nut – 4 off
5	Piston	13	Washer - 4 off
6	Gudgeon pin	14	Small-end needle roller
7	Circlip - 2 off		bearing
8	Expander		

8 Dismantling the engine/gearbox unit: removing the clutch and primary drive pinion

1 The clutch and primary drive pinion can be removed with the engine/gearbox unit in or out of the frame. In either case it will be necessary to remove the right-hand outer casing as described in Section 7 of this Chapter.

2 Slacken and remove the four bolts securing the clutch lifter plate. This must be done progressively by about one turn at a time, in a diagonal sequence to release gradually the pressure of the four clutch springs under the lifter plate. Remove the lifter plate with its integral thrust bearing, the bolts, and the clutch springs.

3 Using a pair of circlip pliers, remove the circlip securing the clutch centre. Remove the clutch centre, the seven clutch plates, and the pressure plate. There should be a splined thrust

washer between the pressure plate and the clutch outer drum. This washer can now be removed from the input shaft together with the clutch outer drum. Keep the washer with the outer drum to assist correct replacement on reassembly.

4 To remove the primary drive pinion the crankshaft must first be locked. If the cylinder head, barrel and piston have already been removed, push a smooth round metal bar through the small-end of the connecting rod and support it on two wooden blocks placed across the crankcase mouth. The metal bar must be a close fit in the connecting rod small-end to prevent damage occurring, and the two wooden blocks are essential to prevent damage to the crankcase mouth when using this method. Alternatively use a strap wrench or similar holding tool to secure the generator rotor. Having locked the crankshaft, slacken and remove the securing nut and slide off the lock washer, primary drive pinion and the spacer behind it. These components should then be put to one side for safekeeping.

8.2a Withdraw clutch lifter plate ...

8.2b ... and the four clutch springs

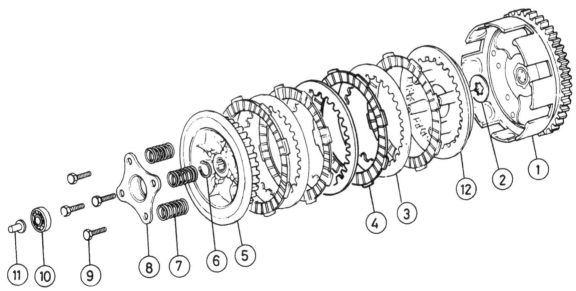

Fig. 1.2 Clutch

1	Outer drum	4	Friction plate - 4 off	7	Spring - 4 off
2	Splined thrust washer	5	Clutch centre	8	Lifter plate
3	Plain plate - 3 off	6	Circlip	9	Bolt - 4 off

10	Thrust bearing
11	Pushrod
12	Pressure plate

9 Dismantling the engine/gearbox unit: removing the balancer idler pinion and oil pump drive pinion

1   Clutch removal is necessary to gain access to the balancer idler pinion and oil pump drive pinion, and is described in full in Section 8 of this Chapter.
2   When the clutch has been removed, pull out the idler shaft complete with its integral double anti-backlash gear pinion, and the oil pump shaft complete with its integral nylon drive pinion. It should be noted that both are only available as complete assemblies and that any attempt at further dismantling would therefore be at least pointless, and at worst the attempt would risk severe engine damage due to the subsequent failure of either of these components.

10 Dismantling the engine/gearbox unit: removing the kickstart shaft assembly and idler gear

1   While the kickstart shaft assembly is easily accessible on removal of the right-hand outer cover, which is described in Section 7 of this Chapter, removal of the clutch is necessary before the kickstart idler gear can be withdrawn. Clutch removal is described in Section 8 of this Chapter.
2   Using a suitable pair of pliers, disengage the hooked arm of the kickstart return spring from its position on the ratchet guide plate, and carefully allow it to unwind until the tension is released. Caution is required here as the spring is under some tension, and could do severe damage if allowed to fly off. The complete kickstart spindle assembly can then be withdrawn and put to one side if it requires no further attention. Do not forget the thrust washer on the outside next to the spring guide or the smaller thrust washer on the inside between the pinion gear and crankcase, These are very easy to lose if not kept with the kickstart assembly.
3   Dismantling of the kickstart shaft assembly need only be done if required. Remove the large thrust washer from the outer end of the shaft and use a suitable pair of pliers to remove the inner end of the kickstart return spring from its locating hole in the kickstart shaft. Withdraw the return spring, the nylon return spring guide, and the light coil spring from the outer end of the shaft, and the smaller thrust washer and the pinion gear from the inner end. Finally slide off the kickstart ratchet. The individual components can then be inspected and replaced as necessary, but should be kept together in a suitable container to prevent their loss or damage until the time comes for re-assembly.
4   Once the clutch and kickstart shaft assemblies have each been removed, the kickstart idler gear can then be withdrawn. Slacken and remove the two bolts which retain the ratchet guide plate and remove the plate, noting its correct position for reassembly. Withdraw the idler gear with the small thrust washer from the end of the output shaft. These components should be kept in a separate container to prevent their loss or confusion with other parts.

11 Dismantling the engine/gearbox unit: removing the gear selector mechanism

1   Prior to removal of the gear selector mechanism the right-hand outer cover, clutch assembly and kickstart assembly must first be removed as described in Sections 7, 8 and 10 of this Chapter.
2   The components of the gear selector mechanism which can be reached outside the main crankcase consist of the gear selector shaft, incorporating the selector claw arm and the pressure spring, the selector shaft return spring, the detent stopper arm, and the gear selector drum complete with its locating pins. These components act on the end of the selector

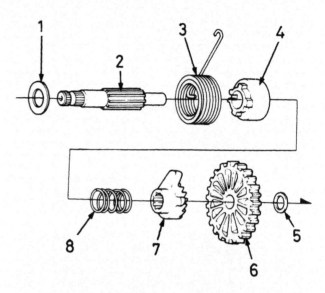

Fig. 1.3 Kickstart components

1	Thrust washer	5	Thrust washer
2	Spindle	6	Kickstart pinion
3	Return spring	7	Ratchet
4	Nylon collar	8	Spring

drum with the stopper arm acting on notches in the camplate fixed to the end of the drum.
3   The detent stopper arm must be removed first by slackening and removing the single bolt which secures it. Care must be taken until the stopper arm spring pressure is released. The camplate securing bolt should now be removed. When removing the camplate the four pins are free to fall out, and should therefore be removed with the plate and put to one side. All four are exactly the same, but are fitted into holes of different depths in the selector drum.
4   The selector shaft can then be withdrawn by easing it out of the main crankcase halves. Take care not to damage the oil seal on the left-hand side as the shaft splines pass through it.

12 Dismantling the engine/gearbox unit: removing the flywheel generator

1   To carry out this operation first release the four bolts securing the left-hand outer cover and remove the bolts and the cover if this has not already been done. Remove the black plastic chain cover piece, if necessary, by unclipping it from its two mounting points on the frame (H100 only).
2   The crankshaft must be locked before the rotor nut can be undone; if the engine/gearbox unit is in the frame when this is to be done, the simplest method is to select top gear and apply the back brake hard, thus locking the crankshaft through the gearbox. The alternatives, if the engine/gearbox unit has been removed from the frame, are either to hold the rotor with a strap wrench or similar holding tool, or to use the method described in Section 8 of this Chapter, ie to use a metal bar through the connecting rod small-end. Whichever method is used, once the crankshaft is locked, slacken and remove the rotor nut and its lock washer.
3   Due to the lack of suitable apertures in the rotor, only one method of rotor removal is possible. This is to use the manufacturer's special tool Part No 07733-0010000, or an equivalent pattern version which is available at many good motorcycle dealers. To use this type of tool, fully unscrew the centre bolt and thread the tool body carefully into the rotor

centre, noting that a left-hand thread is employed. Once the tool body is screwed securely into the rotor as far as possible, tighten the centre bolt with a good quality ring spanner of the correct size, while holding the tool body with another spanner to prevent rotation. When the centre bolt is fully tightened on to the end of the crankshaft, tap the bolt smartly on its head with a hammer. This should immediately shock the rotor free. If it does not, tighten the tool centre bolt further and tap its head again. It should be noted that while this method usually works well at the first attempt, cases have been known of extreme stubbornness on the part of the rotor. If such a case is suspected, take the complete machine or engine/gearbox unit to an authorised Honda dealer for an expert opinion before any damage is done to the crankshaft or rotor due to inexpert or over-enthusiastic use of the rotor removal tool.

4   After removing the rotor, prise the Woodruff key from the tapered portion of the shaft and store it, the rotor, and the rotor nut and lock washer together for safe keeping.

5   As previously stated the apertures in the rotor are not suitable for the use of a two-legged puller, being too small, and too far from the centre of the rotor. Even if the legs of such a tool were small enough and yet strong enough to permit its use, distortion of the rotor would inevitably result. Similarly it is not possible to lever the rotor from the crankshaft as severe damage to the crankcase castings and to the rotor itself could result. It

will therefore, be evident that the rotor can only safely be removed using the proper tool. If this is not available, or it is not wished to purchase one the complete machine or the engine/gearbox unit must be taken to an authorised Honda dealer for the task to be carried out.

6   To remove the stator plate with the engine/gearbox unit still in the frame, trace the generator lead back up to the snap connectors on the right-hand side of the machine and disconnect these and any cable ties which secure the lead to the frame. Remove the cover from the neutral indicator switch and pull out the switch wire which is retained by a spring clip. Slacken and remove the three stator mounting bolts and pull away the stator. It is not necessary to mark the stator.

---

13 Dismantling the engine/gearbox unit: separating the crankcase halves and removing the crankshaft, balancer shaft and gear clusters

---

1   Crankcase separation is necessary to gain access to the crankshaft, balancer shaft, and gearbox components. It can only be carried out after engine removal and preliminary dismantling as described in Sections 4 to 12 of this Chapter.

12.2 Use strap wrench or other methods described to lock rotor during removal

12.3a A special rotor puller must be used to remove the rotor safely

12.3b The special tool in use – see text

12.6 Stator plate is located by dowels – no need to mark it on removal – H100 S

2   Once these preliminary dismantling operations have been carried out, the oil pump must then be removed. Slacken and remove the two securing bolts and lift the pump away, complete with its feed pipes. Store the pump in an upright position to minimise oil leakage and to ease the task of bleeding on rebuilding.

3   Unscrew the eleven crankcase fastening bolts which are all accessible from the left-hand side of the crankcases. Prepare a piece of cardboard as described in Section 7 of this Chapter, with the positions of the bolts marked on it in a clearly identifiable manner. As each bolt is wihdrawn, push it through the cardboard in its correct relative position. Do not forget to place the oil pump cable bracket and breather tube clamps with their respective bolts on the cardboard as shown in the accompanying photograph. This procedure does not take very much time, and greatly eases the task of correctly identifying the bolts on reassembly. Check that all the securing bolts have been removed.

4   The method of crankcase separation recommended by Honda is to use their special tool, Part No 07965-1660000. Bolt the tool outer body in place on the right-hand crankcase with the projecting boss situated centrally over the crankshaft right-hand end. Use three of the right-hand outer cover retaining bolts to secure it. Screw the shorter, plain end of the tool into the tool outer body until it locates against the crankshaft right-hand end. Support the engine/gearbox unit on two wooden blocks on the workbench with the right-hand side uppermost. Apply pressure to the crankshaft by tightening down the tool centre and at the same time tap lightly on the ends of the gearbox shafts and around the crankcase joint area using only a soft-faced mallet. The tapping must be carried out simultaneously with the tightening down of the special tool to ensure that the crankcases separate evenly and with the minimum of strain on the components. Once the crankcases have separated, carry on tightening the tool and tapping the gearbox shaft ends until the right-hand crankcase half has been lifted completely away, leaving the crankshaft and gearbox components in place in the left-hand crankcase half.

5   For those owners who do not have access to the manufacturer's special tool, an alternative method of separating the crankcase halves is given. It should, however, be noted that this latter method requires great care and some skill in its execution if serious damage is not to be done to any of the components. If the owner has any doubts about his ability in this respect, the engine/gearbox unit should be taken to an authorised Honda dealer for the work to be carried out by an expert, on the principle that it is better to pay a small labour charge than a large bill for replacement of needlessly damaged engine/gearbox components. Support the left-hand crankcase on two wooden blocks on the workbench so that the right-hand side is uppermost. Temporarily replace the primary drive pinion retaining nut on the crankshaft right-hand thread, tightening it by hand until the nut is flush with the end of the crankshaft. This will prevent any damage to the thread. Using a soft-faced mallet, tap lightly on the ends of the crankshaft and gearbox shafts to jar the crankcases apart. Also tap the joint area of the crankcase halves to assist separation. Carry on tapping gently until the crankcase halves separate, using the bare minimum of force necessary to achieve this. Once the crankcases have separated, carry on tapping the shaft ends while lifting the right-hand crankcase half with the other hand. Remove the nut on the crankshaft before it fouls the right-hand main bearing. If at any time the crankcases become reluctant to part, stop and find out why. It may be that the right-hand half has lifted unevenly, causing the shafts to stick in their right-hand bearings or that the gasket or a corroded dowel pin is causing the obstruction. If such a problem does occur, gently free the part concerned and carry on. Do not attempt to lever the casings apart with any tool, as this will put a severe strain on a small area of the engine/gearbox unit, causing at the very least damage to the gasket surface of the casting.

6   As soon as the right-hand crankcase half has been lifted away check that there are no thrust washers adhering to its underside, or any other loose components, such as a dowel pin, which might be subsequently lost. Any such part should be replaced immediately in the left-hand crankcase half.

7   Carefully slide out the selector fork shaft and remove the three selector forks. These are all different and should be replaced immediately on the shaft in their correct relative positions to be put to one side. It is a useful aid to correct reassembly to degrease each one as it is removed and to mark it with a spirit-based felt pen in a way that will ensure its correct positioning.

8   Remove the selector drum and balancer shaft, each being easily lifted out of its bearing in the left-hand crankcase.

9   Remove the remaining gearbox components. This is best done holding both shafts together and withdrawing them as a complete unit. It will probably be necessary to use a soft-faced mallet to tap gently on the left-hand end of the output shaft. Ensure that the thrust washers are kept in place on the ends of their respective shafts.

10 Remove the crankshaft. The method recommended by Honda is again the use of their special tool Part No 07965-1660000 which is an extremely useful piece of equipment with several applications. If used, it must be set up on the left-hand crankcase in exactly the same way as previously described for crankcase separation. The short, plain end of the tool is again used, to press on the left-hand crankshaft end. Tighten the tool down carefully and gently, and be careful not to let the crankshaft drop away. If, however, the tool is not available, replace the rotor retaining nut on the left-hand crankshaft end, tightening it by hand only until the nut is flush with the crankshaft end; this is to prevent damage to the thread. Place the crankcase on two wooden blocks so that the left-hand side of the crankcase is uppermost. The two wooden blocks must be situated as close around the crankshaft as possible to give maximum support, and must be of a size to hold the crankcase far enough from the workbench top to allow the crankshaft to be removed. Using a soft-faced mallet with one hand, and supporting the crankshaft with the other, carefully tap the crankshaft out of its housing. Do not use excessive force and do not allow the crankshaft to drop away.

11 If damage has occurred to any part of the gear clusters, or if excessive gearbox wear has taken place, it will be necessary to remove the pinions, with the attendant thrust washers and circlips, from their respective shafts for examination. Always keep the components of the two shaft assemblies separate to avoid confusion on reassembly. Refer to the photographs and drawings in Section 24 for a guide to the correct order of installation of the components.

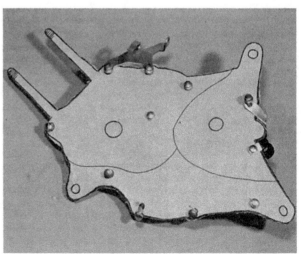

13.3 Method of ensuring that crankcase bolts and fittings are replaced correctly

43

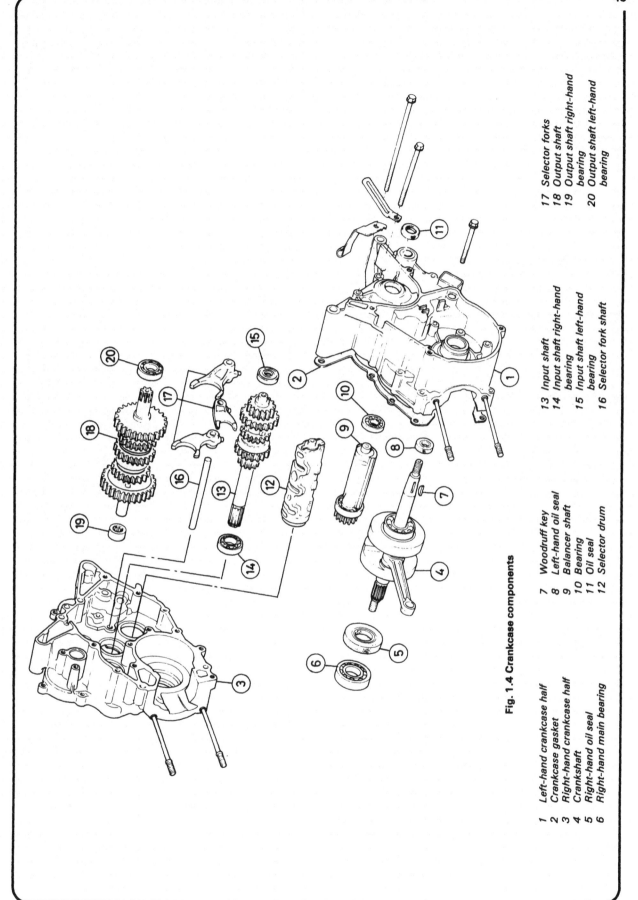

Fig. 1.4 Crankcase components

1 Left-hand crankcase half
2 Crankcase gasket
3 Right-hand crankcase half
4 Crankshaft
5 Right-hand oil seal
6 Right-hand main bearing

7 Woodruff key
8 Left-hand oil seal
9 Balancer shaft
10 Bearing
11 Oil seal
12 Selector drum

13 Input shaft
14 Input shaft right-hand bearing
15 Input shaft left-hand bearing
16 Selector fork shaft

17 Selector forks
18 Output shaft
19 Output shaft right-hand bearing
20 Output shaft left-hand bearing

14 Dismantling the engine/gearbox unit: removal of ancillary components from the crankcase halves

1   If any work is to be done on the crankcase halves themselves, removal of additional items will be required.
2   As a general rule, all the oil seals should be discarded and renewed whenever the crankcases are separated. This is especially true of the crankshaft main bearing oil seals, which are essential to the continued good performance and reliability of the engine. To remove them, a tool must be fabricated from an old broad-bladed screwdriver. Heat the tip with a blowlamp to a point where it can be bent into a slightly curved shape by hammering. Remove any sharp edges with a file. When cooled, insert the tip under one lip of the seal to be removed and lever that side of the seal out, pivoting the screwdriver if necessary on a piece of wood to prevent damage to the casting. Be careful that the screwdriver tip does not scratch the seal housing. Once one side of the seal is levered from its position, it can be easily pulled away by hand. If there is no bearing behind the seal, and if the seal is not retained by a lip in the casting, it can be driven out using a hammer and a socket or tubular drift of suitable size.
3   Press out the neutral indicator switch if necessary. This should be done using the bare minimum of force possible as the switch is only a light plastic moulding.
4   If any of the bearings are to be removed, the casting must first be heated. This is to take advantage of the fact that an aluminium alloy crankcase casting has a higher coefficient of expansion that a steel bearing outer race and that therefore the casting will, when heated, expand faster than the bearings. This will have the effect of loosening the bearings in their housings to the point where they can relatively easily be drifted or pulled away. To achieve this, the complete crankcase half must be heated to approximately 100°C by a gradual application of heat over its entire surface to prevent the distortion which would result from a fierce local application of heat. An oven is the best method of heating the casting, but an alternative is to place the casting in a suitable container and then to pour boiling water over it. Do not use a welding torch or blowlamp for this operation as the even application of heat cannot be guaranteed by the inexperienced. It will be evident that, whichever method is used, great care must be taken both with the method of heating and in subsequently handling the heated casting, if serious personal injury is to be avoided.
5   Bearings which are to be used again must be removed by tapping them out of their housings using a hammer and a socket spanner or tubular drift which bears only on the outer race of the bearing concerned. If any other part of the bearing

is used, unacceptably high side loadings will be placed on the balls or rollers and their cages, causing premature failure. If the entire outer race is not accessible, for example where a bearing is fitted in a blind housing, some other means of bearing removal will have to be found. The most widely used method is to prepare a clean, flat wooden surface, heat the casting as previously described and tap the casting firmly and squarely on to the wooden surface with enough force to jar the bearings free. Great care must be taken that the casting is tapped squarely on to a clean surface to avoid damaging the gasket surface. An alternative method is to heat the casting as described, rest it on a clean wooden surface to support it evenly, and to tap the casting with a soft-faced mallet directly behind the bearing, again with the object of jarring the bearing free. It is stressed that in both cases, tapping must be as gently as possible to avoid damaging the casting. If either method fails, take the casting to an authorised Honda dealer for the bearings to be removed using a slide hammer with an internal puller atrachment.

15 Examination and renovation: general

1   Before examining the parts of the dismantled engine unit for wear it is essential that they should be cleaned thoroughly.

14.2 Lever oil seals carefully from their housings

14.3 Only disturb the neutral indicator switch if necessary

14.4 Casting must be heated to permit bearing removal

Use a petrol/paraffin mix or a high flash-point solvent to remove all traces of old oil and sludge which may have accumulated within the engine. Where petrol is included in the cleaning agent normal fire precautions should be taken and cleaning should be carried out in a well ventilated place.

2 Examine the crankcase castings for cracks or other signs of damage. If a crack is discovered it will require a specialist repair.

3 Examine carefully each part to determine the extent of wear, checking with the tolerance figures listed in the Specifications Section of this Chapter or in the main text. If there is any doubt about the condition of a particular component, play safe and renew.

4 Use a clean lint free rag for cleaning and drying the various components. This will obviate the risk of small particles obstructing the internal oilways, and causing the lubrication system to fail.

5 Various instruments for measuring wear are required, including a vernier gauge or external micrometer and a set of standard feeler gauges. Both an internal and external micrometer will be required to check wear limits. Additionally, although not absolutely necessary, a dial gauge and mounting bracket is invaluable for accurate measurement of end float, and play between components of very low diameter bores – where a micrometer cannot reach.

After some experience has been gained the state of wear of many components can be determined visually or by feel, and thus a decision on their suitability for continued service can be made without resorting to direct measurement.

16 Examination and renovation: crankcase and fittings

1 The crankcase halves should be thoroughly degreased, using one of the proprietary water-soluble degreasing solutions such as Gunk. When clean and dry a careful examination should be made, looking for signs of cracks or other damage. Any such fault will probably require either professional repair or renewal of the crankcases as a pair. Note that any damage around the various bearing bosses will normally indicate that crankcase renewal is necessary. because a small discrepancy in these areas can result in serious mis-alignment of the shaft concerned. It is important to check crackcase condition at the earliest opportunity, because this will permit remedial action to be taken and any necessary machining or welding to be done whilst attention is turned to the remaining engine parts.

2 In most cases, badly worn or damaged threads can be reclaimed by fitting a thread insert. This is a simple and inexpensive task, but one which requires the correct taps and fitting tools. It follows that the various threads should be checked and the cases taken to a local engineering works or motorcycle dealer offering this service so that repair can take place while the remaining engine parts are checked.

17 Examination and renovation: crankshaft and main bearings

1 Check the crankshaft assembly visually for damage, paying particular attention to the slot for the Woodruff Key and to the threads at each end of the mainshaft. Should these have become damaged specialist help will be needed to reclaim them.

2 The connecting rod should be checked for big-end bearing play. A small amount of end float is normal, but any up and down movement will necessitate renewal. Grasp the connecting rod and pull it firmly up and down. Any movement will soon become evident, but be careful that axial clearance (endfloat), a set amount of which is permissible, is not mistaken for wear. Assuming that the big-end bearing is in good order, attention should be turned to the rest of the connecting rod. Visually check the rod for straightness, particularly if the engine is being

rebuilt after a seizure or other catastrophe. Look also for signs of cracking. This is extremely unlikely, but worthwhile checking. Spotting a hairline crack at this stage may save the engine an an untimely end. Should any wear or damage be found, it will be necessary to consult an expert for advice as to whether repair is possible or if the purchase of a new crankshaft assembly is advisable.

3 If measuring facilities are available, set the crankshaft in V-blocks and check the big-end radial clearance using a dial gauge mounted on a suitable stand. Check crankshaft run-out in the same way. Big-end axial clearance (end float) is measured with feeler gauges of the correct thickness which should be a firm sliding fit between the thrust washer next to the big-end eye and the machined shoulder on the flywheel. If any clearance exceeds the wear limits given in the specifications Section of this Chapter the crankshaft assembly should be taken to an authorised Honda dealer or similar repair agent for repair or renewal.

4 It should be noted that crankshaft repair work is of a highly specialised nature and requires the use of equipment and skills not likely to be available to the average private owner. Such work should not be attempted by anyone without this equipment and the skill to use it.

5 A caged roller bearing is employed as the small-end bearing. This bearing can be removed quite easily for examination. Check the rollers for any imperfection, renewing the bearing if less than perfect. The gudgeon pin and small-end eye should be checked where the rollers bear upon them, and remedial action taken where the surface(s) are marked. Assemble the small-end bearing and gudgeon pin in the connecting rod eye, and check for radial play. If any movement is found, renew the bearing, particularly if the engine has produced a characteristic rattle in the past, indicating that all might not be well with this bearing.

6 The crankshaft main bearings are of the ball journal type and usually remain in place in their respective crankcase halves on removal of the crankshaft. In this case, if they are to be removed for examination, the removal procedure is as described in Section 14 of this Chapter. If, however, they stick on the crankshaft as it is removed, a conventional knife-edged bearing puller must be used to renew them for examination or renewal. They should be washed thoroughly in clean petrol (gasoline) and checked for radial and axial play. Spinning the bearing when dry will highlight any rough spots, producing obviously excessive amounts of noise once the lubricating film has been removed. Any signs of pitting or scoring of the bearing tracks or balls indicates the need for renewal. If there is any doubt about the condition of the main bearings, they should be renewed.

17.2 Crankshaft repair work should be entrusted only to a specialist

17.6 Main bearings should be examined and renewed if necessary

18 Examination and renovation: piston and piston rings

1   It cannot be over-emphasised that the condition of the piston and piston rings is of prime inportance because they control the opening and closing of the ports by providing an effective moving seal. A two-stroke engine has only three working parts, of which the piston is one. It follows that the efficiency of the engine is very dependent on the condition of the piston and the parts with which it is closely associated.

2   Remove the piston rings by pushing the ends apart with the thumbs whilst gently easing each ring from its groove. Great care is necessary throughout this operation as the rings are brittle and will break easily if overstressed. If the rings are gummed in their grooves, three strips of tin can be used to ease them free as shown in the accompanying illustration. The expander ring in the lower ring groove should then be removed. This is less brittle, but is very thin and should be picked out with great care. It should be noted that if it has been decided that a rebore is necessary, due to obvious and serious wear or damage in the bore or piston and rings, no further attention will be required to the piston and rings as they will be replaced with oversized components.

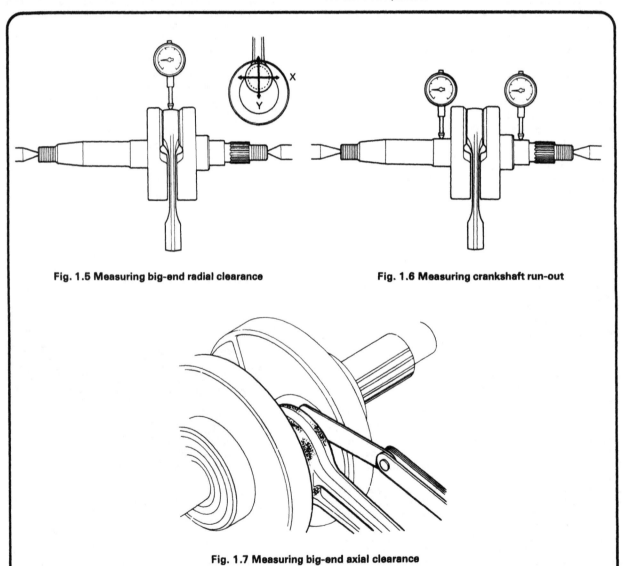

Fig. 1.5 Measuring big-end radial clearance

Fig. 1.6 Measuring crankshaft run-out

Fig. 1.7 Measuring big-end axial clearance

3   If a rebore is not yet considered necessary, the piston must be examined closely. Remove all carbon from the piston crown using a soft metal or hard wooden scraper to avoid scratching the piston. Finish off with metal polish to obtain a highly polished finish as carbon will adhere much less readily to such a surface. Examination will show if the cylinder barrel has previously been rebored as the amount of overbore is invariably stamped in the piston crown. This figure, if present, must be borne in mind when subsequently measuring the piston and cylinder bore, the stated amount being added to the set of figures for the overall diameter of the piston or the internal diameter of the bore as necessary. Remove all traces of carbon from the piston ring grooves, using a length of broken piston ring to avoid enlarging the grooves. Measure the internal diameter of the gudgeon pin bore and the overall diameter of the gudgeon pin itself. Compare these figures with the wear limits set. If no measuring equipment is available, the gudgeon pin/piston clearance can be assumed to be good if the gudgeon pin is a tight push fit in the piston and if there is no discernible play when the gudgeon pin is in the installed position. If the measurements or the practical check reveal excessive wear, the piston or gudgeon pin must be renewed as necessary. Examine the piston ring grooves. Unfortunately no set figure for piston ring/ring groove side clearance is given by the manufacturer. Assessment of this aspect of piston wear is, therefore, largely a matter of experience. Check that the ring upper and lower surfaces are not worn, noting that they are of the keystone type which has a smoothly tapered upper surface as standard. If in doubt, purchase a new set of rings and fit them. If the piston ring up and down movement appears excessive, take the piston to an authorised Honda dealer for an expert opinion. This point is important because if excessive clearance is not corrected on rebuilding, the piston rings will be overstrained by the fluttering action imposed on them as the engine is running, and will eventually break. This will almost certainly cause enough damage to the cylinder bore to make reboring or even cylinder barrel renewal necessary. If the piston has passed these checks, its overall diameter must now be measured to determine the amount of wear that has taken place. Details of this measurement are given in Section 19, as it is an important factor when determining by measurement if the cylinder barrel needs reboring.

4   The piston rings should now be examined. The working surfaces must be clean and polished throughout, any discolouration showing that the rings have not been sealing against the bore surface. The upper and lower sides must be clean and unworn and the inside surface free from carbon. If they pass this test, insert them one at a time into an unworn bore, using the piston crown to push them squarely down the bore to a depth of approximately $1\frac{1}{2}$ inches from the top of the bore. Using feeler gauges of appropriate thickness, measure the piston ring end gap. If the figure measured is greater than that given in the Specifications Section, the piston rings must be renewed. It should be noted that many people renew the piston rings as a matter of course to ensure that maximum performance is restored. Although this decision must be left to the owner, it is recommended that if there is any doubt about the condition of the rings they should be renewed.

5   If new rings are to be fitted the end gap should be checked as just described. Do not assume that new components will automatically fit your machine perfectly as even the most carefully made parts will have some variation due to manufacturing tolerances. The cylinder bore must be very lightly honed to remove the glazed finish and provide a slightly roughened surface which will enable the new piston rings to bed in properly. If this is not done, the rings will not be able to bed in as well or as fast as they should, giving a poor seal, and subsequent poor performance. The honing operation must be done by a motorcycle dealer with the necessary equipment.

6   Check that the piston ring pegs are firmly embedded in each piston ring groove. It is imperative that these retainers should not work loose, otherwise the rings will be free to rotate and there is a danger of the ends being trapped in the ports.

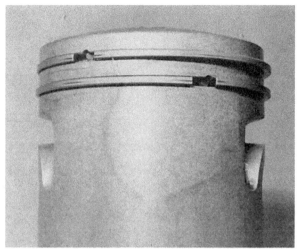

18.1 It is essential that the condition of piston and rings is carefully checked

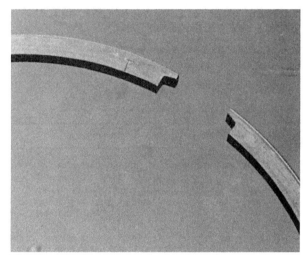

18.4 Carefully clean and examine the piston rings

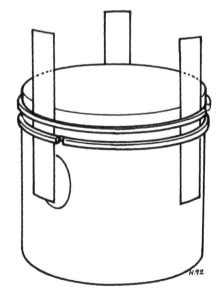

Fig. 1.8 Method of removing and replacing piston rings

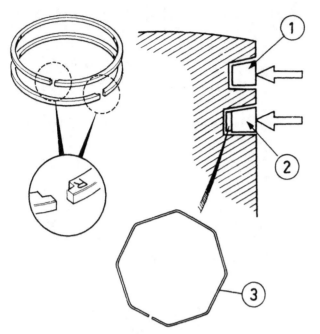

Fig. 1.9 Piston ring profiles

1   *Top ring*
2   *Second ring*
3   *Expander ring*

19 Examination and renovation: cylinder barrel

1   The cylinder barrel should be carefully cleaned using a wire brush and petrol to remove any accumulation of grime around the cooling fins. After drying the bore with a clean rag, examine the surface for signs of wear or scoring. If scoring or scratches are in evidence in the bore, the cylinder will need to be rebored and a new piston fitted.

2   A small ridge may be in evidence near to the top of the bore. This marks the extent of travel of the top piston ring, and will probably be more pronounced at one point (the thrust face) than at any other. If this is barely perceptible, and the piston and rings are in good condition, it will probably be safe to use the existing bore. If in any doubt, and in any case if the ridge is marked, the barrel should be taken to a Honda dealer for checking, together with its piston.

3   For those owners who have the correct equipment the condition of the piston and cylinder barrel may be determined by direct measurement. Measure the overall diameter of the piston skirt at a point 10 mm ($\frac{3}{8}$ in) from the base of the skirt and at right angles to the gudgeon pin axis. The figure measured should be noted and compared with the wear limit given in the Specifications Section of this Chapter. If less than the wear limit specified, the piston must be renewed. The cylinder bore should be measured at a point 15 mm ($\frac{5}{8}$ in) below the top edge of the cylinder both along the axis of the gudgeon pin and at right angles to it. Similar measurements should be made in the middle and at the bottom of the bore, avoiding the port areas. The manufacturer specifies that the minimum figure thus noted be taken to determine the amount of bore wear. Subtract the overall diameter of the piston from this figure. If the piston/cylinder clearance figure thus obtained is greater than the wear limit given in the Specifications Section, the cylinder barrel is in need of reboring. It should be noted that if the piston was found to be excessively worn, as determined by measurement, the overall diameter of a new piston given in the Specifications Section must be substituted for subtraction from the bore internal diameter. If the piston/cylinder clearance then

comes within the set tolerances and the cylinder bore is otherwise undamaged it will suffice to purchase a new piston of the same size, whether it be standard or oversized, as necessary. Remember that due allowance must be made in calculating the above clearances, if the cylinder barrel has been rebored already as previously mentioned.

4   Reboring should be carried out by an authorised Honda dealer or similar repair agent who will be able to supply the necessary oversized piston and piston rings.

5   Clean all carbon deposits from the exhaust ports using a blunt ended scraper. It is important that all the ports should have a clean, smooth appearance because this will have the dual benefit of improving gas flow and making it less easy for carbon to adhere in the future. Finish off with metal polish, to heighten the polishing effect.

6   Do not under any circumstances enlarge or alter the shape of the ports under the mistaken belief that improved performance will result. The size and position of the ports predetermines the characteristics of the engine and unwarranted tampering can produce very adverse effects.

19.1 If cylinder bore is excessively worn, reboring will be necessary

20 Examination and renovation: cylinder head

1   It is unlikely that the cylinder head will require any special attention apart from removing the carbon deposits from the combustion chamber. Finish off with metal polish; the polished surface will help improve gas flow and reduce the tendency of future carbon deposits to adhere so easily.

2   Check that the cooling fins are clean and unobstructed, so that they receive the full air flow.

3   Check the condition of the thread within the sparking plug hole. The thread is easily damaged if the sparking plug is overtightened. If necessary, a damaged thread can be reclaimed, by fitting a Helicoil thread insert. Most Honda dealers have facilities for this type of repair, which is not expensive.

4   If there has been evidence of oil seepage from the cylinder head joint when the machine was in use, check whether the cylinder head is distorted by laying it on a sheet of plate glass. Severe distortion will necessitate renewal of the cylinder head, but if distortion is only slight, the head can be reclaimed by wrapping a sheet of emery paper around the glass and using it as the surface on which to rub down the head with a rotary motion, until it is once again flat. The usual cause of distortion is failure to tighten down the cylinder head nuts evenly, in a diagonal sequence.

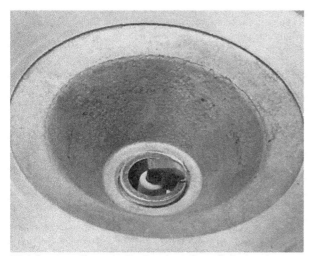

20.1 Remove all traces of carbon from the cylinder head combustion chamber

21 Examination and renovation: gearbox components

1   The gearbox bearings should be examined in situ to ensure that they are disturbed as little as possible. Wash each bearing in a petrol/paraffin mixture and allow it to dry. Spinning the bearing as fast as possible will highlight any rough spots, as without a lubricating film a worn bearing will produce obviously excessive amounts of noise. Any signs of pitting or scoring of the bearing tracks, balls, or rollers, or any other damage revealed by close inspection will indicate the need for renewal of the bearing concerned. Any bearing which is to be renewed must be removed and the new bearing fitted in accordance with the directions given in Section 14 of this Chapter.

2   The gearbox oil seals should be renewed irrespective of their condition. Once they have been disturbed they should be removed as described in Section 14 of this Chapter, and discarded.

3   Examine the gear selector fork shaft for signs of damage and roll it on a sheet of plate glass or a similar flat surface. This test will immediately reveal if it is bent or distorted. If measuring equipment is available, compare the measured outside diameter of the shaft with that given in the Specifications Section of this Chapter. If it is worn at any point beyond the set limit, it must be renewed. If no measuring equipment is available, push its entire length through each of the selector forks in turn. If the shaft is worn, it will only be on those portions of its surface which carry a selector fork and any wear will be immediately apparent due to the normal push fit suddenly becoming excessively sloppy. Equally, this test will reveal wear, if present, in the bores of the selector forks should the correct measuring equipment not be available for these. Again, excessive wear can only be corrected by renewing the parts concerned. If the shaft is merely bent, it can be straightened, but this task should only be done by an expert.

4   The selector forks should be examined closely to ensure that they are not bent or badly worn. If measuring equipment is available, check the dimensions measured against those given in the Specifications Section. Note that the bores through which the selector fork shaft passes can be checked, if measuring equipment is not available, as described above. Selector fork wear is restricted normally to the fork claw ends and will be readily apparent, especially if the fork is bent, due to the blueing of the claw end which is caused by constant excessive pressure on the rotating gear pinions. If trouble has been experienced with gear selection or jumping out of gear, the selector forks should be very carefully examined. Any sign of wear or damage can only be corrected by renewing the part concerned.

5   If measuring equipment is available, carefully measure the outside diameters of the two bearing surfaces of the selector drum and compare the figures taken with those given in the Specifications Section. Refer to the accompanying illustration to ensure that measurements are made at the correct place. The tracks in the selector drum should be checked for signs of excessive wear, but it should be noted that this is extremely rare unless neglect has led to under-lubrication of the gearbox. Also check the gearchange shaft, claw arm, and drum stopper arm springs for tension. Any weakness in these springs will contribute towards imprecise gear selection. Carefully examine the gearchange shaft assembly, drum stopper arm, camplate and the selector drum/camplate pins. Wear in any of these items will mean that a replacement part should be purchased before reassembly.

6   Examine each of the gear pinions to ensure that there are no chipped or broken teeth and that the dogs on the end of the pinions are not rounded. Where such figures are given, measure the internal diameter of the gear pinions and compare the measurements obtained with those given in the Specifications Section. Gear pinions with any such defects or excessive wear must be renewed; there is no satisfactory method of reclaiming them.

7   The gearbox shafts should be examined carefully. If the shafts are bent on their splines and circlip grooves are badly worn, they must be renewed. Refer to the accompanying illustration which shows in detail at which points the outside diameters of the shafts are to be measured. These are the points at which the gear pinions rotate on the shaft instead of being fixed by splines. Compare the figures taken with those given in the Specifications Section, and renew any shaft or corresponding gear pinion which is worn beyond the set limit. Similarly measure very carefully the inside and outside diameters of the brush which supports the output shaft 2nd gear pinion. If it is worn beyond the wear limit for either dimension the bush should be renewed. Excessive clearance between the shafts and their corresponding gear pinions will produce a disproportionate level of noise and will promote rapid and severe gearbox wear if not corrected during rebuilding.

8   Refer to Section 24 and the accompanying illustrations and photographic sequences when fitting the gear pinions to the shafts and note that it is advisable to use new thrust washers and circlips throughout. The new parts should be obtained with the new seals and gaskets required for reassembly.

9   Check the condition of the kickstart components. If slipping has been encountered a worn ratchet and pawl will invariably be traced as the cause. Any other damage or wear to the components will be self-evident. If either the ratchet or pawl is found to be faulty, both components must be replaced as a pair. Examine the kickstart return spring which should be renewed if there is any doubt about its condition.

21.1 Remove and examine gearbox bearings as described in the text

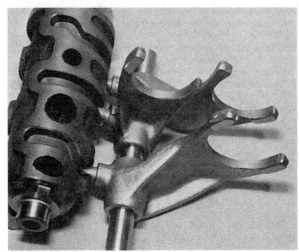

21.3 Carefully check gear selection components for wear or damage

Fig. 1.10 Selector drum bearing surfaces measurement

A At 13 mm diam. 12.934 – 12.984 mm (0.5092 – 0.5112 in)
Wear limit 12.850 mm (0.5059 in)
B At 36 mm diam. 35.950 – 35.975 mm (1.4154 – 1.4163 in)
Wear limit 35.900 mm (1.4134 in)

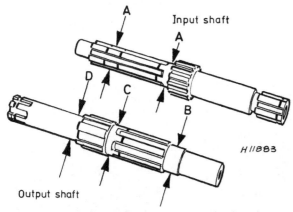

Fig. 1.11 Gearbox shaft wear measurement locations

A Wear limit 16.93 mm (0.6665 in)
B Wear limit 16.44 mm (0.6472 in)
C Wear limit 18.93 mm (0.7453 in)
D Wear limit 16.96 mm (0.6677 in)

---

22 Examination and renovation: clutch assembly

---

1   After an extended period of service the clutch friction plates will wear and promote clutch slip. Each plate should be measured for thickness and the measurement compared with the wear limit given in the Specifications Section. When the wear limit is reached, the plates must be renewed, preferably as a complete set.

2   The three plain plates should not show any excess heating (blueing). Check the warpage of each plate using plate glass or a surface plate and a feeler gauge. The maximum warpage allowed is given in the Specifications Section.
3   The clutch springs will lose tension after a period of use, and should be renewed as a precaution if clutch slip has been evident and the friction plates are within limits. The free length of the clutch springs gives a good indication of condition, and this should be checked and compared with the figures given in the Specifications Section.
4   Examine the clutch assembly for burrs or indentations on the edges of the protruding tongues of the friction plates and/or slots worn in the edges of the outer drum with which they engage. Similar wear can occur between the inner tongues of the plain clutch plates and the slots in the clutch inner drum. Wear of this nature will cause clutch drag and slow disengagement during gear changes, since the parts will become trapped and will not free fully when the clutch is withdrawn. A small amount of wear can be corrected by dressing with a fine file; more extensive wear will necessitate renewal of the worn parts.
5   The clutch release mechanism takes the form of a spindle running in the right-hand outer casing, the shaped end of which bears on the clutch release pushrod when the handlebar lever is operated. The mechanism is of robust construction and requires no attention during normal maintenance or overhauls.

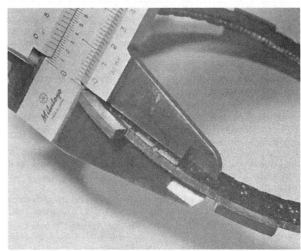

22.1 Measure each friction plate to assess amount of wear

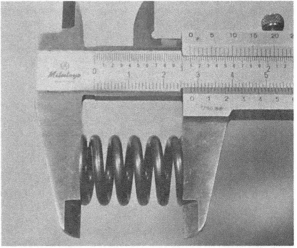

22.3 Measure clutch spring free length and renew springs if necessary

22.5 Clutch release mechanism should require little attention

23 Engine reassembly: general

1   Before reassembly of the engine/gear unit is commenced, the various component parts should be cleaned thoroughly and placed on a sheet of clean paper, close to the working area.
2   Make sure all traces of old gaskets have been removed and that the mating surfaces are clean and undamaged. Great care should be taken when removing old gasket compound not to damage the mating surface. Most gasket compounds can be softened using a suitable solvent such as methylated spirits, acetone or cellulose thinner. The type of solvent required will depend on the type of compound used. Gasket compound of the non-hardening type can be removed using a soft brass-wire brush of the type used for cleaning suede shoes. A considerable amount of scrubbing can take place without fear of harming the mating surfaces. Some difficulty may be encountered when attempting to remove gaskets of the self-vulcanising type, the use of which is becoming widespread, particularly as cylinder head and base gaskets. The gasket should be pared from the mating surface using a scalpel or a small chisel with a finely honed edge. Do not, however, resort to scraping with a sharp instrument unless necessary.
3   Gather together all the necessary tools and have available an oil can filled with clean engine oil. Make sure that all new gaskets and oil seals are to hand, also all replacement parts required. Nothing is more frustrating than having to stop in the

middle of a reassembly sequence because a vital gasket or replacement has been overlooked. As a general rule each moving engine component should be lubricated thoroughly as it is fitted into position.
4   Make sure that the reassembly area is clean and that there is adequate working space. Refer to the torque and clearance setting wherever they are given. Many of the smaller bolts are easily sheared if overtightened.

24 Engine reassembly: rebuilding the gearbox clusters

1   If the gearbox components have been dismantled for examination and renewal, it is essential that they are rebuilt in the correct order to ensure proper operation of the gearbox. Use the accompanying photographic sequence and line drawing as aids to the identification of components and their correct relative positions.
2   Take the input shaft with its integral 1st gear and slide the plain thrust washer over its left-hand end to butt up against the 1st gear pinion. This is followed by the 4th gear pinion, which is fitted with the selector dogs facing the left-hand end of the shaft. Secure the 4th gear pinion by fitting first a splined thrust washer and then a circlip.
3   Next is the 3rd gear pinion, which is fitted with its selector fork groove inwards, adjoining the 4th gear. Secure this with a circlip. Slide another splined thrust washer along the shaft until it butts against the circlip. The 5th gear pinion is then fitted with its drilled face inwards to mate with the selector dogs on the 3rd gear pinion. Last is the 2nd gear pinion followed by a thick spacer. Put the completed input shaft assembly to one side.
4   Turning to the output shaft, slide the 3rd gear pinion on to its right-hand end with its selector dog holes facing outwards. Secure it with a splined thrust washer and circlip. Slide on the 4th gear pinion with its selector fork groove towards the 3rd gear. Follow this with the 1st gear pinion, which is fitted with its recessed side inwards, and the large, plain, thrust washer.
5   Turning to the left-hand end of the shaft, first slide on the 5th gear pinion with its selector fork groove inwards. Then fit a plain thrust washer and the plain bush which supports the 2nd gear pinion. Ensure that both surfaces of the bush are well lubricated before fitting the next item, which is the 2nd gear pinion itself. The 2nd gear pinion must be fitted with its recessed face inwards. Lastly fit the thick spacer over the left-hand end of the then completed output shaft assembly.
6   If this has not already been done, ensure that all bearing surfaces on the shafts and gear pinions are well lubricated. Check that the new circlips are properly located in their grooves and that the pinions slide or rotate smoothly and easily where applicable. Put the completed gearbox cluster to one side to await reassembly.

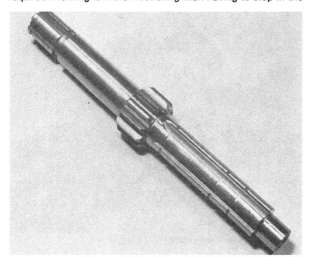

24.2a Take the bare input shaft with its integral 1st gear ...

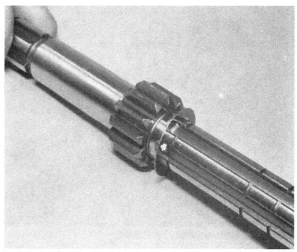

24.2b ... and slide on the plain thrust washer ...

24.2c ... followed by the 4th gear pinion which is retained by ...

24.2d ... a splined thrust washer ...

24.2e ... and a circlip

24.3a Next is the 3rd gear pinion ...

24.3b ... which is retained by a circlip ...

24.3c ... followed by a splined thrust washer ...

24.3d ... and the 5th gear pinion, against which is fitted ...

24.3e ... the 2nd gear pinion, followed by ...

24.3f ... a thick spacer

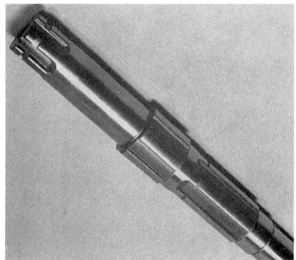
24.4a Take the bare output shaft ...

24.4b ... slide the 3rd gear pinion and a splined thrust washer into place ...

24.4c ... over the shaft right-hand end. Secure with a circlip ...

24.4d ... and fit the 4th gear pinion ...

24.4e ... followed by the 1st gear pinion ...

24.4f ... and the large plain thrust washer

24.5a Slide 5th gear pinion over the shaft left-hand end ...

24.5b ... followed by a plain thrust washer ...

24.5c ... and the plain bush, which should be well lubricated ...

24.5d ... before the 2nd gear pinion is fitted

24.5e The thick spacer fits against the 2nd gear pinion

24.6 The completed gearbox shaft assemblies

25 Engine reassembly: fitting the crankshaft

1   The left-hand crankcase half should be completely clean and dry at this stage. If any of the gearbox bearings are to be replaced in the crankcase, this must be done now. Heat the crankcase to approximately 100°C as described in Section 14 of this Chapter, and gently tap the bearing squarely into place using a hammer and a tubular drift or socket spanner. As also described in Section 14, the drift must bear on the bearing outer race only.

2   If the crankshaft left-hand main bearing is still in position in the crankcase it need not be disturbed for crankshaft refitting. If, however, the bearing has been removed for examination or a new one is to be fitted, it should be fitted first to the crankshaft and the two items should then be fitted in the crankcase as a single unit. This will ensure that the bearing is positioned correctly. To fit the bearing, first warm it slightly in an oven, then tap it into place on the crankshaft left-hand end using a hammer and a tubular metal drift which bears on the inner race of the bearing. Be careful to support only the left-hand flywheel

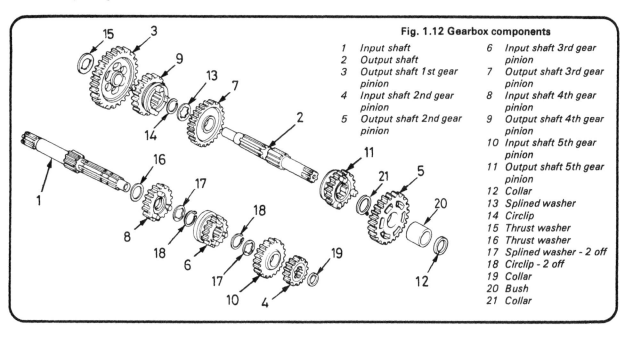

Fig. 1.12 Gearbox components

1   Input shaft
2   Output shaft
3   Output shaft 1st gear pinion
4   Input shaft 2nd gear pinion
5   Output shaft 2nd gear pinion
6   Input shaft 3rd gear pinion
7   Output shaft 3rd gear pinion
8   Input shaft 4th gear pinion
9   Output shaft 4th gear pinion
10 Input shaft 5th gear pinion
11 Output shaft 5th gear pinion
12 Collar
13 Splined washer
14 Circlip
15 Thrust washer
16 Thrust washer
17 Splined washer - 2 off
18 Circlip - 2 off
19 Collar
20 Bush
21 Collar

of the crankshaft in this operation, to prevent the shock forces being passed through the crankpin and possibly causing crankshaft misalignment.

3   The recommended method of fitting the crankshaft is to use the manufacturer's special tool, Part No 07965–1660000. Screw the tool centre adaptor as far as possible up the thread of the tool centre and push it through the tool outer body until the adaptor locates firmly on the shoulder of the projecting boss. Heat the crankcase casting in an oven to approximately 100°C. If the left-hand main bearing is in place on the crankshaft, it should now be lubricated liberally, both around its balls and their tracks, and on the outside diameter of the outer track. Pass the special tool through the crankcase from its left-hand side and screw the threaded, hollow, end of the tool centre on to the crankshaft left-hand end. Tighten the tool centre clockwise, as a left-hand thread is employed, until the crankshaft or its main bearing is butted up against the crankcase. Bolt the tool outer body securely to the crankcase to ensure that the crankshaft is pulled squarely into place, and make a last check that the crankshaft is correctly located. Gently tighten the tool centre to draw the crankshaft into position. If the casing was heated and the bearing lubricated as described, the crankshaft should slide easily into place with very little resistance until the main bearing locates against the crankcase wall. If undue resistance is encountered, remove the tool and find out why. Do not under any circumstances force the crankshaft using this tool as serious damage could be done if the tool is improperly used.

4   If, as is likely, the special tool is not available the following method can be used to fit the crankshaft, but great care must be taken. Heat the crankcase casting as described above. Replace the primary drive pinion nut on the crankshaft right-hand end, tightening the nut by hand only until it is flush with the crankshaft end. If the left-hand main bearing is in position in the crankcase, lightly lubricate the crankshaft left-hand shaft; if, however, the main bearing is to be installed with the crankshaft, liberally lubricate its outside diameter. Support the heated crankcase half on two wooden blocks placed side by side immediately under the main bearing housing, but with enough distance between the two blocks to allow the crankshaft left-hand shaft to protrude beneath the casting when it is fitted. Place the crankshaft in position, and being very careful to hold it exactly vertically to the crankcase, tap it into place using a soft-faced mallet on the crankshaft right-hand end. The primary drive pinion nut will protect the thread from damage. It must be stressed that great care must be taken to follow the instructions

exactly, which will make the task as easy as possible, and that the bare minimum of force necessary is used.

5   Whichever method is used, once the crankshaft is fully located in its housing, lubricate the left-hand main bearing and big-end bearing with two-stroke oil, and check that the crankshaft assembly revolves easily and freely. Remove the nut, if used, from the crankshaft right-hand end.

26 Engine reassembly: fitting the gearbox components

1   With the left-hand crankcase half lying on two wooden blocks to support it, lubricate the gearbox bearings and bearing surfaces. If the neutral indicator switch was removed, it should now be pressed gently back into place, using a new O-ring to prevent oil leakage.

2   Lubricate both bearings of the balancer shaft and replace it in its left-hand bearing in the crankcase. Check that it revolves easily and freely.

3   Take the two gearbox shaft assemblies and match them together with the gears in their correct relative positions. Carefully slide them into their bearings in the crankcase ensuring that the thick spacers on the left-hand end of each shaft do not fall clear. It may be necessary to tap the right-hand end of the output shaft while supporting the cluster with the other hand. Lightly lubricate the bearing surfaces of the shafts and their respective pinions and ensure that they revolve freely and that the sliding gear pinions are free to move.

4   Check that the neutral indicator switch blade is correctly fitted on the left-hand end of the selector drum and fit the drum into the left-hand crankcase. Rotate the drum to line up the switch blade with the indicator switch itself.

5   Replace the selector forks. Lightly oil each one before installation. The left-hand one operates on the output shaft 5th gear pinion, the centre one operates on the input shaft 3rd gear pinion, and the right-hand one operates on the output shaft 4th gear pinion. Lightly oil the selector fork shaft and slide it through the forks into its housing in the left-hand crankcase.

6   Carefully check that the bearing surfaces of any moving parts are properly lubricated and ensure that all four shafts are quite free to rotate. Check carefully that no parts have been omitted and ensure that the two locating dowel pins are firmly located in the left-hand crankcase. Lightly smear a new crankcase centre gasket with grease and position it on the left-hand crankcase.

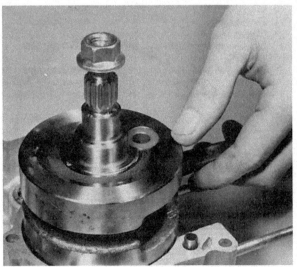

25.4 Carefully fit the crankshaft in the left-hand crankcase half

26.2 Refit the balancer shaft in its bearing

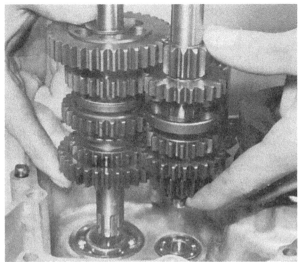

26.3 Gearbox shaft assemblies should be fitted as one unit

26.4a Ensure that the switch contact is correctly fitted ...

26.4b ... and install the selector drum

26.5a Install first the left-hand selector fork ...

26.5b ... then the centre fork ...

26.5c ... and the right-hand fork

26.5d Lubricate the selector fork shaft and push it into place

27 Engine reassembly: fitting the right-hand crankcase half

1   If the crankshaft right-hand main bearing was removed for examination, or if a new component is to be fitted, this must now be done.

2   Heat the right-hand crankcase to approximately 100°C as described in Section 14. It is recommended that an oven is used in this instance, to ensure that the casting remains dry during reassembly. Once the casting is heated, support it on a clean, flat wooden surface and tap the bearing firmly into place using a hammer and a tubular drift or socket spanner which bears only on the outer race of the bearing to avoid damaging it. Ensure that the bearing is kept absolutely square as it is tapped in. Grease the outside diameter and inner sealing lip of a new oil seal and tap this into place using a hammer and a tubular drift or socket spanner which fits on its outside diameter only to prevent distortion of the seal. Liberally lubricate the main bearing and gearbox bearings using engine oil.

3   The recommended method of fitting the right-hand crankcase is to use the manufacturer's special tool, Part No 07965-1660000 mounted on the crankshaft right-hand end and used in exactly the same way as described in Section 25 of this Chapter when fitting the crankshaft. Care must be taken to ensure that the crankcase halves stay perfectly square while the tool is used. Tapping the crankcase around the gearbox shafts with a soft-faced mallet while tightening the tool will help to achieve correct crankcase alignment.

4   The alternative is to lower the right-hand crankcase half into position over the engine and gearbox shaft ends and to tap it all the way into position using a soft-faced mallet. When using this method be careful to preserve correct crankshaft alignment. If any resistance is encountered stop tapping and check all the shafts to ensure that they are in correct alignment. Also ensure that the two locating dowel pins are correctly aligned. Be careful at all times to use only the bare minimum of force necessary to complete the operation.

5   Once the right-hand crankcase half is firmly in position, check that all the shafts are free to rotate easily. Rotate the selector drum to assess gearbox selection. If any doubts arise from these checks the problem should be cured before any more assembly work is done; if necessary, by separating the crankcases again to investigate.

6   When it is established that the crankshaft and gearbox are properly installed, invert the crankcase assembly so that the left-hand side is now uppermost. Replace the crankcase securing bolts in their correct positions, remembering to include the oil pump cable adjuster bracket and breather tube clamps, and tighten them down. Start in the middle, working outwards in a diagonal sequence, and tighten the bolts in two stages to ensure an even and progressive application of pressure.

7   Lubricate the left-hand main bearing wth two-stroke oil and use a hammer and a suitably-sized tubular drift to tap a new crankcase oil seal into position. Grease applied to the inner and outer edges of the seal will ease the fitting operation and help prevent damage to the delicate sealing lips. Similarly fit the output shaft and gearchange shaft oil seals. Remember that the tubular drift used must bear only on the hard outer diameter of the oil seal being fitted to prevent distortion of the seal.

8   Ensure that the gearbox drain plug washer is in good condition, renewing it if necessary and replace the drain plug, using a torque wrench to tighten it to 2.0 – 2.5 kgf m (14 – 18 lbf ft). Pack the crankcase mouth with clean rag to prevent dirt getting in during reassembly.

27.2a Use socket to drift in right-hand bearing

27.2b Fit new main bearing seals as a matter of course

27.4a Ensure that all bearings are well lubricated ...

27.4b ... and use a new gasket (note two locating dowel pins) ...

27.4c ... before replacing the right-hand crankcase half

27.7a Use plenty of grease to aid fitting of left-hand main bearing ...

27.7b ... and gearbox output shaft oil seals

28 Engine reassembly: fitting the flywheel generator

1   Fit the generator stator with the line scribed next to one of the mounting bolt holes aligned exactly with the raised index mark on the crankcase wall (H100 only). On H100 S models refit the stator plate ensuring that the wiring is correctly routed, then press the dowel pins in to place to locate accurately the plate. Refit and tighten securely the three mounting bolts. Using a suitably-sized pair of pliers, press down on the neutral indicator switch terminal and fit the switch wire through the hole thus exposed. Replace the switch cover and ensure that the wire is located in the retaining lugs in the crankcase and that the rubber grommet on the generator lead is seated correctly in the recess provided for it in the crankcase top.

2   Tap the Woodruff key into position in the crankshaft keyway and carefully replace the rotor. Replace the nut and lock washer. Lock the crankshaft as described in Section 12 of this Chapter and tighten the rotor nut to 6.0 – 7.0 kgf m (43 – 50 lbf ft).

28.1a Replace stator plate ensuring that wiring is correctly routed ...

28.1b ... and that timing marks (arrowed) are aligned exactly

28.1c Connect neutral indicator switch wire ...

28.1d ... and replace indicator switch cover

28.2a Replace Woodruff key in crankshaft keyway ...

28.2b ... and install generator rotor

28.2c Crankshaft must be locked to allow rotor nut to be tightened to correct torque setting

29 Engine reassembly: fitting the gear selector mechanism

1   Lightly oil the gear selector shaft and slide it through the crankcase halves. Take care not to damage the oil seal on the left-hand side as the gear lever splines pass through it. Support the engine/gearbox unit on two wooden blocks with the right-hand side uppermost. Slide the selector shaft fully into position ensuring that the selector claw arm fits next to the selector drum and that the selector shaft return spring is correctly engaged on its locating peg.
2   Using a suitable pair of pliers, fit the four camplate locating pins in their holes in the end of the selector drum. Engage the selector claw arm against the bottom pin. Fit the camplate, noting that the two raised pins fit into two holes in its underside. Apply some thread locking cement to the securing bolt, which should then be tightened down.
3   Apply thread locking cement to the detent stopper arm bolt and fit the stopper arm assembly in position. Screw the bolt down lightly then ensure that the spring is located correctly with its hooked end in the groove on the stopper arm and with its straight end butting against the crankcase bottom wall. Push the stopper arm against spring pressure and engage the roller

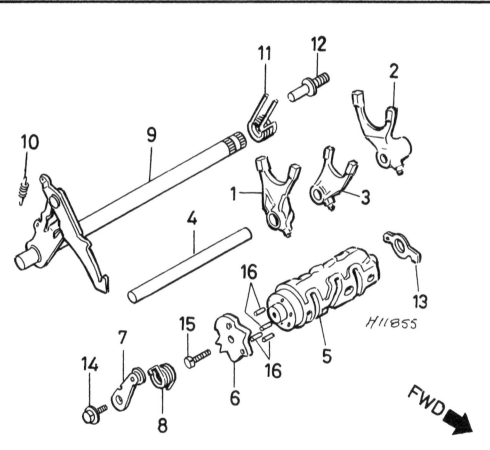

Fig. 1.13 Gearchange mechanism

1	Right-hand selector fork	7	Detent stopper arm	12	Locating peg
2	Left-hand selector fork	8	Return spring	13	Neutral switch contact
3	Centre selector fork	9	Gearchange shaft	14	Bolt
4	Selector fork shaft	10	Claw arm return spring	15	Bolt
5	Selector drum	11	Shaft return spring	16	Pin - 4 off
6	Camplate				

into its notch in the camplate. Tighten down the securing bolt.
4   Temporarily replace the gearchange lever on the gear selector shaft and check that all gears can be selected easily and that there is no undue stiffness in the gear selector mechanism. Note that it will probably be necessary to rotate the input shaft in its normal direction of rotation, ie anti-clockwise, to faciliate this check.

30 Engine reassembly: fitting the kickstart mechanism

1   Place the thrust washer over the end of the output shaft, lightly lubricate the shaft,and place the idler gear in position. Apply thread locking cement to the two bolts securing the ratchet guide plate, place the guide plate in position and tighten down the two bolts.
2   Slide the kickstart pinion gear on to the kickstart spindle with the ratchet teeth facing the splined end of the spindle. Fit

the smaller thrust washer behind it and locate the spindle in its machined boss in the crankcase. Check that both gears are free to rotate.
3   Slide the kickstart ratchet over the kickstart spindle so that the punch mark on the ratchet lines up with the spring hole in the spindle. Then fit the light coil spring and the nylon collar. Turn the spindle so that the ratchet arm locates against the ratchet guide plate.
4   Fit the kickstart return spring with its inner end located in the spring hole in the spindle, and with the notch in the nylon collar located around the spring inner end. Using a suitable pair of pliers tension the kickstart return spring by moving the long hooked end clockwise until it can be engaged on the ratchet guide plate. Be careful to support the kickstart spindle assembly with the other hand during this operation. Fit the large thrust washer over the end of the spindle.
5   Temporarily replace the kickstart lever on the kickstart spindle and operate it gently to check that all is free to move and working properly. If all is well, remove the lever.

29.1 Ensure that selector shaft return spring is correctly engaged

29.2a Replace camplate locating pins and engage selector claw arm on bottom pin

29.2b Use thread locking cement on camplate retaining bolt

29.3 Detent stopper arm assembly in correct position

30.1a Fit plain thrust washer over end of output shaft ...

30.1b ... followed by the idler gear ...

30.1c ... which is retained by fitting the ratchet guide plate

30.2a The kickstart spindle thrust washer is fitted against the crankcase ...

30.2b ... followed by the kickstart pinion gear ...

30.2c ... and the kickstart spindle

30.3a Align the punch mark (arrowed) on the ratchet with the spring hole in the kickstart spindle ...

30.3b ... and replace the light coil spring ...

30.3c ... followed by the nylon spring guide

30.4a Inner end of kickstart return spring must locate in the spindle and in the notch in the spring guide

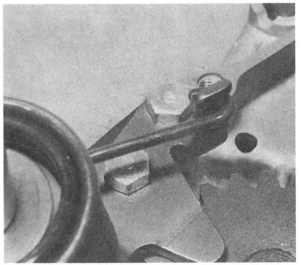

30.4b Fit large plain thrust washer ...

30.4c ... and engage long hooked spring end on ratchet guide plate

31 Engine reassembly: fitting the primary drive gear and oil pump

1   Lock the crankshaft as described in Section 8 of this Chapter. Fit the primary drive pinion spacer on to the right-hand end of the crankshaft. This is followed by the pinion itself which has a punch mark on its outer face. Carefully align this exactly with the punch mark in the crankshaft end. Fit the lock washer and nut. Tighten the nut down to 4.5 – 5.5 kgf m (33 – 40 lbf ft).

2   Lightly oil the balancer idler gear and slide it into position. Make sure that both sets of the inner pair of gears line up properly with the teeth on the balancer shaft gear.

3   Check the condition of the O-ring on the oil pump spigot and renew the O-ring if necessary. Lightly grease the O-ring to aid fitting and slide the oil pump into place. Tighten down the two mounting bolts. Lightly lubricate the oil pump drive gear and install it, aligning the slot in the drive gear shaft with the blade in the pump. Locate the oil pump feed pipe in its bracket on the crankcase.

31.1a Fit first the spacer over crankshaft end

31.1b Then replace the primary drive pinion so that ...

31.1c ... the timing marks (arrowed) align exactly

31.1d Replace lock washer and nut

31.1e Note method used to lock crankshaft to allow nut to be tightened

31.2a Install balancer idler gear assembly and ensure ...

31.2b ... that both sets of inner pair of gears are engaged

31.3a Grease oil pump spigot and O-ring to aid assembly

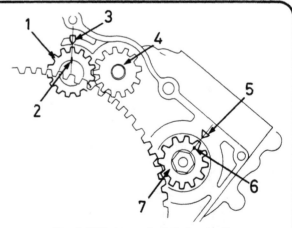

31.3b Install oil pump drive gear ensuring that drive is correctly engaged

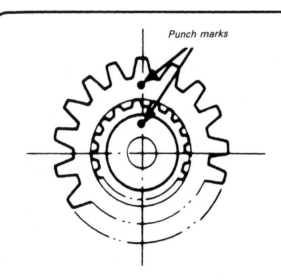

Punch marks

Fig. 1.14 Primary drive pinion alignment punch marks

Fig. 1.15 Balance shaft timing marks

1	Balance shaft gear pinion	5	Crankcase index mark
2	Alignment mark	6	Punch mark
3	Crankcase index mark	7	Primary drive pinion
4	Balancer idler gear		

32 Engine reassembly: fitting the clutch assembly

1 Carefully align the marks on the primary drive pinion and balancer shaft gear with their respective index marks on the crankcase. Refer to the accompanying illustration which shows this.

2 Lightly oil the bush in the clutch outer drum and slide the clutch outer drum into position on the input shaft. Patience and care will be required at this point as the alignment of the balancer timing marks must not be disturbed and both sets of the outer pair of gears on the balancer idler shaft must be correctly engaged on the clutch outer drum teeth. Ensure that the timing marks are aligned and that the clutch outer drum is fully in position before proceeding. Do not use any force whatsoever in fitting the clutch outer drum. When it is in place fit the splined thrust washer with its rounded surfaces facing outwards.

3 For ease of assembly the clutch plates should be fitted on to the clutch centre before installation. Start with a friction plate and continue assembly, fitting plain and friction plates alternately. Note that where new friction plates are fitted, these should be coated with gearbox oil before assembly. Fit the clutch pressure plate, passing the four projecting pillars through the corresponding holes in the clutch centre. Carefully align the projecting tongues on the friction plates. This is important to ensure that the assembly will then slide easily into the clutch outer drum. Secure the clutch pressure plate/clutch centre assembly with a new circlip.

4 Place the four clutch springs in place over their corresponding pillars and then fit the lifter plate, complete with its ball bearing, tightening the four bolts by hand alone at first. Complete the tightening by using a spanner at about one turn at a time and in a diagonal sequence. This ensures the progressive and even application of spring pressure.

32.1 Ensure that balancer timing marks (arrowed) are aligned exactly

32.2a Great care is required when fitting clutch outer drum (see text)

32.2b Thrust washer is fitted with its rounded surface outwards

32.3a It is easiest to assemble clutch plates on to clutch centre before fitting in to clutch outer drum

32.3b Always use a new circlip to secure clutch centre

32.4 Tighten clutch lifter plate bolts progressively and evenly

33 Engine reassembly: fitting the right-hand outer cover

1   Check that the two locating dowel pins are in position on the crankcase. Lightly smear a new cover gasket with grease and place it in position. Grease the kickstart spindle splines.
2   Check that the clutch lifting mechanism is in place in the right-hand outer cover, using grease to stick it in position if necessary. Grease the inner sealing lips of the kickstart spindle oil seal to minimise damage to them. Carefully lower the right-hand outer cover into position, taking care not to damage the kickstart spindle oil seal as it passes over the shaft splines. A few careful taps using a soft-faced mallet may be necessary to seat the cover. Note that force should not be necessary in this operation, and if any obstruction is encountered, the cover should be removed again to investigate the cause.
3   Once the cover is firmly in position, check that the clutch operating arm is free to move and then fit the ten hexagon-headed screws in their corresponding positions, not forgetting the clutch cable bracket. Tighten the bolts down in a diagonal sequence starting from the centre. Check the operation of the clutch arm again.
4   Refill the gearbox with the correct amount of gearbox oil. Remember to check the level after the engine has first been run. Check the tightness of the filler cap and drain plug.

34 Engine reassembly: fitting the piston, cylinder barrel and cylinder head

1   Fit the piston rings to the piston. The thin expander ring fits in the lower ring groove. The two compression rings are interchangeable but must be fitted the correct way up. It will be seen that there is a small mark, usually the letter 'T', stamped or etched on one surface of each ring in the vicinity of the ring gap. This mark must face towards the top of the piston. Check that the ring gaps are correctly aligned with the locating peg in each groove on the piston.
2   Check that the clean rag is still in place in the crankcase mouth. Oil the small-end bearing and place it in the connecting rod. Position the piston over the connecting rod with the 'IN'

marking facing the inlet side ie, to the rear of the engine. The gudgeon pin should be lightly oiled and then pushed through the piston and small-end bearing. If it is difficult to do this, warm the piston by soaking a rag in hot water, wringing it out, and wrapping it round the piston. Secure the gudgeon pin with two new circlips and ensure that these are seated firmly in their grooves.
3   Lightly grease a new cylinder base gasket and slide it over the cylinder studs into position on the crankcase mouth. It should be noted that due to the shape of the ports and the offset position of the studs, there is only one way this gasket will fit properly. Check this carefully before damage is done to the gasket. Bring the piston up to TDC.
4   Lubricate liberally the cylinder bore, piston, and rings with two stroke oil. Carefully place the barrel on the studs and lower it down on to the piston. Due to the low weight and small size of the parts concerned, this operation should be completed easily by one person. Compress the top piston ring by hand and gently push the barrel down over it. The lead in or chamfer at the bottom of the bore makes this a relatively easy task. If there is any difficulty check that the piston is entering squarely into the bore and that the ring gaps are correctly located at their respective pegs. Repeat the procedure with the second ring. When both rings have engaged the bore, remove the rag from the crankcase mouth. Push the barrel gently down to rest firmly on the crankcase.
5   Lightly grease a new head gasket and place it in position on the cylinder, noting that the raised sealing lips must face up. Lower the cylinder head into place and hand tighten the four sleeve nuts. Using a torque wrench, tighten the four nuts to 1.9 – 2.3 kgf m (14 – 17 lbf ft). Tighten the nuts in a diagonal sequence and in two stages to ensure a progressive and even application of pressure. Do not overtighten these nuts.
6   Lightly grease the new reed valve gaskets and fit the reed valve assembly to the cylinder barrel, followed by the inlet stub. Fit and tighten down the four securing bolts. The bolts should be tightened in a diagonal sequence but not overtightened, otherwise the delicate inlet stub and reed valve castings will be distorted.
7   Fit the spark plug, ensuring that it is clean and correctly gapped, and tighten it down, by hand only, to prevent the entry of dirt. Temporarily unplug the end of the oil feed pipe and push it over the union on the inlet stub.

33.1 Place a new gasket over two locating dowels (arrowed)

33.2 Carefully replace right-hand outer cover

33.3 Note correct position of clutch cable bracket

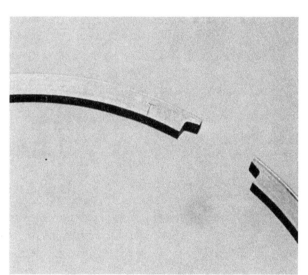

34.1 Install piston ring with etched letter upwards

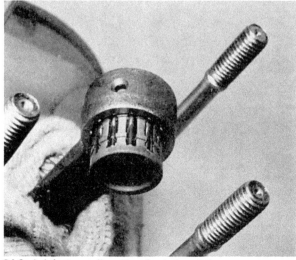

34.2a Lubricate small-end bearing before fitting

34.2b 'IN' marking faces to rear of engine

34.2c Gudgeon pin should be a tight press fit in piston

34.2d Always use new circlips to secure gudgeon pin

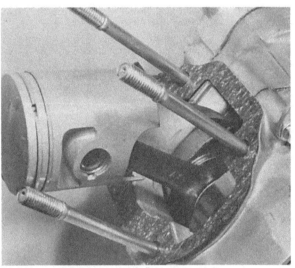

34.3 New cylinder base gasket fits one way only

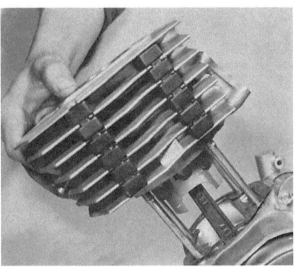

34.4 Remove rag when both piston rings are in cylinder bore

34.5a Note correct position of cylinder head gasket

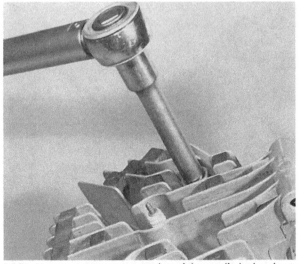

34.5b Always use a torque wrench to tighten cylinder head nuts evenly

34.6a Fit new reed valve gaskets to prevent induction leaks

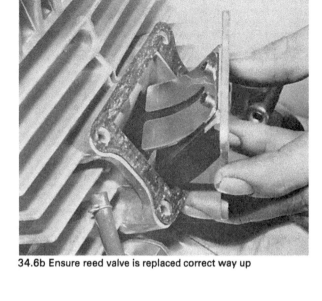

34.6b Ensure reed valve is replaced correct way up

34.6c Inlet stub is delicate and must be handled carefully

35 Installing the rebuilt engine in the frame: final adjustments

1   Lift the engine/gearbox unit up until the top mounting on the cylinder head lines up with the bracket on the frame. Push the top mounting bolt through from left to right. Replace the nut but only tighten it by hand for the moment.

2   Swing the bottom half of the engine gently backwards into position, guiding the carburettor on to the inlet stub mounting studs with the other hand. Ensure that the carburettor drain tube and battery breather pipe are located in the tube clamps on the rear of the crankcase and are not trapped against the frame. Slide the upper rear mounting bolt into position, locating a spacer on each side of the engine/gearbox unit (H100 only). Similarly, slide in the engine bottom mounting bolt. Hand tighten their respective nuts. Check to ensure that the engine/gearbox unit is correctly seated in its mountings and that nothing is trapped. Using a torque wrench, tighten the three engine mounting bolt securing nuts to 3.0 – 4.0 kgf m (22 – 29 lbf ft). Carefully replace and tighten down the two carburettor retaining nuts.

3   Engage the gearbox sprocket on the chain and slide the sprocket over the end of the output shaft. Fit the sprocket retaining plate next to the sprocket and turn it until the retaining bolt holes are aligned. Fit and tighten down the two bolts, applying the back brake to lock the rear wheel if necessary. Check, and adjust if necessary, the final drive chain tension and rear brake adjustment

4   Unplug the oil tank/oil pump feed pipe and the oil tank union, and replace the feed pipe in the union. Secure the wire clip. Note that if the feed pipe was securely plugged after disconnection, it will still be full of oil. If however the pipe has been disturbed and is partially or completely drained, it would be best to free the wire clip and disconnect it at its lower, oil pump, end and to allow the oil to drain through it from the tank until all air bubbles have been expelled. Once the oil tank/oil pump feed line is completely free from air the pipe should be connected to the oil pump again and the wire clip secured. If this operation is done at this stage, it is easier due to the better access available, and will ease the task of bleeding air from the oil injection system which must be carried out before the engine is run for any length of time. Once the oil feed line is free of air and its wire clips have been secured, the oil pump cable can be installed. Fit the cable end nipple into the control lever on the oil pump body and fit the cable adjuster in the bracket provided on the crankcase top for this purpose. Adjust the cable as described in Section 18 of Chapter 2 and check that both the oil pump lever and throttle slide are operating smoothly and fully.

5   Fit the footrest bar and tighten down the three mounting bolts to 2.5 – 3.5 kgf m (18 – 25 lbf ft).

6   Moving round to the right-hand side of the machine, lightly grease a new exhaust gasket and place it in the exhaust port. Manoeuvre the exhaust pipe into position and hand tighten the two front flange securing nuts. Replace and hand tighten the rear mounting bolts on the frame. Once the exhaust pipe is properly mounted, fully tighten the first two nuts and then the bolts.

7   Fit the clutch cable end nipple into the clutch operating lever and locate the cable adjuster in the bracket provided for this purpose. Turn the adjuster nut as necessary to give 10 – 20 mm ($\frac{3}{8}$ – $\frac{3}{4}$ in) free play at the top of the clutch handlebar lever. Replace the kickstart lever on its spindle using the marks made on removal as an aid to correct positioning. Tighten the pinch bolt to 0.8 – 1.2 kgf m (6 – 9 lbf ft).

8   Pass the generator lead up through the frame tubes to the main loom, and connect the individual wires at their snap connectors. Use the colour coding of the wires as a guide.

Secure the lead using any clamps or cable ties provided for this purpose and replace the rubber waterproofing sleeve over the snap connectors. Replace the spark plug cap on the spark plug. Replace the battery in its holder and secure the battery retaining strap. Connect the two battery leads at their snap connectors. Replace the side panel on its mountings and tighten the retaining screw or catch.

9   If this has not already been done, fill the gearbox with 1.0 litre (1.76 Imp pint) of oil but do not bother checking the level until the machine has been run for a while, before it is taken out on the road. Some time must be allowed for the oil to be distributed around the freshly rebuilt components and for an accurate oil level check to be made.

---

36 Starting and running the rebuilt engine

---

1   Turn on the fuel tap, close the choke, and attempt to start the engine by means of the kickstart pedal. Do not be disillusioned if there is no sign of life initially. A certain amount of perseverance may prove necessary to coax the engine into activity even if new parts have not been fitted. Should the engine persist in not starting, check that the spark plug has not become fouled by the oil used during re-assembly. Failing this go through the fault finding charts and work out what the problem is methodically.

2   When the engine does start, keep it running as slowly as possible to allow the oil to circulate. Open the choke as soon as the engine will run without it. During the initial running, a certain amount of smoke may be in evidence due to the oil used in the reassembly sequence being burnt away. The resulting smoke should gradually subside.

3   Once the engine has warmed up enough to tick over smoothly, stop it and carry out the complete oil bleeding procedure as detailed in Section 19 of Chapter 2. It is absolutely essential that all traces of air are removed from the oil injection system before the system is run further.

4   As soon as bleeding has been carried out, so that the engine is receiving the correct amount of lubricant and can therefore be run safely at a higher speed, the ignition timing must be checked using a timing lamp. This is to ensure that the stator plate was replaced in the correct position on reassembly and need be done only if the stator was disturbed. The procedure is given in full in Section 9 of Chapter 3 or in Routine

maintenance. When the ignition timing has been checked, ensure that the neutral indicator switch cover is still in place and fit the left-hand outer cover and black plastic cover. Tighten the four securing bolts. Replace the gearchange lever using the marks made on removal as an aid to correct positioning. Tighten the pinch bolt to 0.8 – 1.2 kgf m (6 – 9 lbf ft).

5   When the bleeding operation and ignition timing check have been carried out, the engine should have been run for long enough for the oil level in the gearbox to have settled. Stop the engine, and with the machine placed on its centre stand on level ground, remove the gearbox level plug. Oil should merely trickle from the orifice. If too much comes out, allow oil to drain until the flow is reduced. If not enough oil appears, add a small amount through the filler plug orifice. When the level is correct, replace the level and filler plugs and tighten them securely. Check that the drain plug has been tightened to a torque setting of 2.0 – 2.5 kgf m (14 – 18 lbf ft). Wash away any surplus oil from the exterior of the engine.

6   Check the engine for blowing gaskets and oil leaks. Before using the machine on the road, check that all gears select properly, and that the controls function correctly.

35.1 Allow engine/gearbox unit to hang in frame

35.2a Check routing of breather pipes and wiring ...

35.2b ... before engine is swung into position and mounting bolts ...

35.2c ... are replaced. Tighten nuts to ...

35.2d ... recommended torque settings (see text)

35.3 Replace gearbox sprocket and adjust chain if necessary

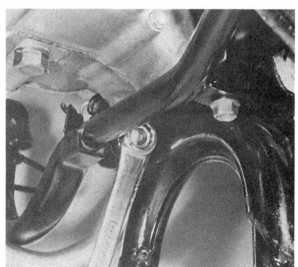

35.5 Footrest bar is retained by three bolts – H100

35.6a Always use a new exhaust gasket ...

35.6b ... and tighten two front mounting nuts first

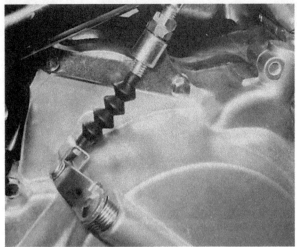

35.7a Clutch cable is adjusted at adjuster on cable bracket

35.7b Ensure kickstart lever is correctly positioned before tightening pinch bolt

35.9 Check gearbox oil level, topping up if necessary

36.4a Black plastic cover is clipped on to frame tubes – H100

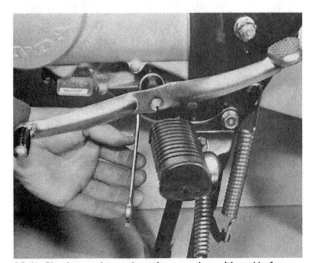

36.4b Check gearchange lever is correctly positioned before tightening pinch bolt

37 Taking the rebuilt machine on the road

1   Any rebuilt machine will need time to settle down, even if parts have been replaced in their original order. For this reason it is highly advisable to treat the machine gently for the first few miles to ensure oil has circulated throughout the lubrication system and that new parts fitted have begun to bed down.

2   Even greater care is necessary if the engine has been rebored or if a new crankshaft has been fitted. In the case of a rebore, the engine will have to be run in again, as if the machine were new. This means greater use of the gearbox and a restraining hand on the throttle until at least 500 miles have been covered. There is no point in keeping to any set speed limit; the main requirement is to keep a light loading on the engine and to gradually work up performance until the 500 mile mark is reached. These recommendations can be lessened to an extent when only a new crankshaft is fitted. Experience is the best guide since it is easy to tell when the engine is running freely.

3   Remember that a good seal between the piston and the cylinder barrel is essential for the correct functioning of the engine. A rebored two-stroke engine will require more careful

running-in, over a long period, than its four-stroke counterpart. There is a far greater risk of engine seizure during the first hundred miles if the engine is permitted to work hard.

4   If at any time a lubrication failure is suspected, stop the engine immediately and investigate the cause. If an engine is run without oil, even for a short period, irreparable engine damage is inevitable.

5   Do not on any account add oil to the petrol under the mistaken belief that a little extra oil will improve the engine lubrication. Apart from creating excess smoke, the addition of oil will make the mixture much weaker, with the consequent risk of overheating and engine seizure. The oil pump alone should provide full engine lubrication.

6   Do not tamper with the exhaust system or run the engine without the baffle fitted to the silencer. Unwarranted changes in the exhaust system will have a marked effect on engine performances, invariably for the worse. The same advice applies to dispensing with the air cleaner or the air cleaner element.

7   When the initial run has been completed allow the engine unit to cool and then check all the fittings and fasteners for security. Re-adjust any controls which may have settled down during initial use.

Chapter 2 Fuel system and lubrication

Contents

Specifications

Fuel capacity

	H100	H100 S
Overall	10 lit (2.2 gal)	11 lit (2.4 gal)
Reserve	2.4 lit (0.53 gal)	2.0 lit (0.44 gal)

Carburettor

Make	Keihin
Type	PF25A
Venturi diameter	18 mm (0.71 in)
Main jet	115 (alternative fitment H100 S only – 125)
Main air jet	150
Needle clip position – grooves from top	2nd
Pilot jet – H100 S	42
Float height	13.5 mm (0.53 in)
Pilot air screw – turns out	$1\frac{1}{8}$
Idle speed	1300 rpm

Air filter

Type	Oiled polyurethane foam

Engine lubrication

Type	Honda 2-stroke oil injection system

Oil tank capacity:

H100	1.6 lit (2.8 pint)
H100 S	1.2 lit (2.1 pint)

Gearbox oil capacity

	H100	H100 S
At engine rebuild	1.00 lit (1.76 pint)	0.95 lit (1.67 pint)
At oil change	0.90 lit (1.58 pint)	0.85 lit (1.50 pint)

1 General description

The fuel system comprises a petrol tank from which fuel is fed via a tap and pipe to a throttle valve type Keihin carburettor. A plunger type choke is fitted to aid cold starting. To prevent the ingress of dust and other foreign matter an air filter is fitted. The polyurethane foam element is oiled for maximum filtering efficiency and is housed in a plastic filter box on the left-hand side of the machine. The filter box is connected directly to the carburettor.

The engine is lubricated by Honda's own two-stroke oil injection system and the gearbox components and primary drive by splash from oil contained in a reservoir formed by the crankcase castings. Both methods of lubrication are described in greater detail in the relevant Sections.

2 Petrol/oil tank: removal and refitting

1 On H100 models, use pliers to disengage the wire retaining clips on the petrol tank breather pipe, oil tank/oil pump feed

line, and the petrol pipe. Slide the clips down their respective pipes until they are clear of the mounting spigots. Turn the petrol tap to the 'Off' position and remove the petrol tank breather pipe and petrol pipe by pulling them gently off their respective mounting spigots. Similarly, disconnect the oil tank/oil pump feed line but place a finger over the outlet from the oil tank, temporarily to stop the flow of oil. Both oil pipes must then be plugged using screws or bolts of suitable size to prevent the loss of oil from the tank and the entry of dirt or air into the oil tank/oil pump feed line. Care must be taken not to damage the pipe ends.

2 Once the various pipes have been disconnected, slacken and remove the two bolts situated on the underneath of each side of the seat, at the rear. They also retain the rear flashing indicator lamp assemblies. Raise the rear of the seat and disengage it from its front mounting which is a hook engaged under a bracket on the frame. This will expose the single petrol tank rear mounting bolt which must then be slackened and removed. Raise the rear of the petrol tank and slide it backwards off its two front mounting rubbers.

3 Refitting the petrol/oil tank is a straightforward reversal of the removal procedure. When the tank and seat have been refitted and their respective mounting bolts securely tightened,

unplug first the oil tank/oil pump feed line and then the oil tank outlet. Allow a small amount of oil to drain into the feed line, displacing any air that may be present in the feed line. Push it back on to the oil tank union and secure the wire retaining clip, then wipe off any surplus oil. Check for oil leaks and also for fuel leaks after the petrol pipe has been refitted and its clips secured. Also check carefully for signs of air in the oil tank/oil pump feed line. Any air bubbles must be removed by bleeding as described in Section 19.

4   The procedure for removing and refitting the petrol tank on H100 S models is as described above, except that there is no oil feed pipe or breather pipe to disconnect. If it is necessary to remove the oil tank at any time, remove the right-hand side panel and swing out the tank, as described in Routine maintenance, then disconnect the feed pipe as described above before detaching the tank from its mountings. Observe the same precautions to ensure that all air bubbles are removed on reassembly.

2.1a Disconnect oil tank/oil pump feed line at union directly beneath petrol/oil tank – H100

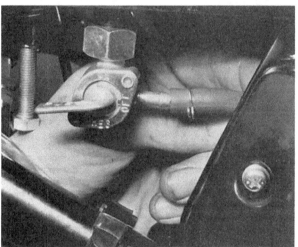

2.1b Switch petrol tap to 'Off' position before disconnecting petrol pipe

2.2a Seat is retained by two bolts at the rear ...

2.2b ... and by a hook at the front

2.2c Seat removal exposes the petrol tank rear mounting bolt

2.2d Petrol/oil tank must be lifted up at the rear and pulled backwards off its front mountings

3   Petrol/oil tank: examination and renovation

1   On H100 models, the two-stroke oil supply for the oil injection system is contained in a separate compartment within the main petrol tank. If any work is to be done to the oil compartment, it must first be very carefully drained and all traces of oil removed by swilling clean petrol around inside it. Where applicable, remove the gauge unit by very carefully unclipping the plastic retaining ring from the tank itself. If any faults occur in the gauge unit it should be examined. Depending on the nature of the fault, some repairs may be possible, but it should be noted that the gauge is only available as a complete assembly and that no parts are available to recondition it.

2   Inspect the whole tank assembly for signs of petrol leakage or rusting, both of which will be immediately apparent, and will require attention. If traces of dirt have been appearing continually in the fuel line and carburettor or in the oil tank/oil pump feed line, the tank must be very carefully washed out, rinsed in clean petrol and checked for serious internal rusting.

3   If the tank is to be stored it should be placed in a safe place away from any area where fire is a hazard or where the paint finish may become damaged.

4   Any sign of fuel leakage should be dealt with promptly in view of the risk of fire or explosion should fuel drip onto the hot exhaust system. It is not recommended that the tank is repaired using welding or brazing techniques, because even a small amount of residual fuel vapour can result in a dangerous explosion. A more satisfactory alternative is to use one of the resin-based tank sealing compounds. These are designed to line the tank with a tough fuel-proof skin, sealing small holes, or splits in the process. The suppliers of these products advertise regularly in the motorcycle press.

4   Petrol feed pipe: examination

1   The petrol feed pipe is made from thin walled synthetic rubber and is of the push-on type. It is necessary to replace the pipe only if it becomes hard or splits. It is unlikely that the retaining clips will need replacing due to fatigue as the main seal between the pipe and union is effected by an interference fit.

2   If the petrol pipe has been replaced with the transparent plastic type for any reason, look for signs of yellowing which indicates that the pipe is becoming brittle due to the plasticiser being leached out by the petrol. It is a sound precaution to renew a pipe when this occurs, as any subsequent breakage whilst in use will be almost impossible to repair. **Note:** On no account should natural rubber tubing be used to carry petrol, even as a temporary measure. The petrol will dissolve the inner wall, causing blockages in the carburettor jets which will prove very difficult to remove.

5   Petrol tap: examination and refitting

1   Before the petrol tap can be removed, it is first necessary to drain the tank. This is easily accomplished by removing the feed pipe from the carburettor float chamber and allowing the contents of the tank to drain into a clean receptacle, with the tap turned to the 'Reserve' position. Alternatively, the tank can be removed and placed on one side, so that the fuel level is below the tap outlet. Take care not to damage the paintwork.

2   The tap unit is retained by a gland nut to the threaded stub on the underside of the tank. It can be removed after the fuel pipe has been pulled off the tap.

3   If the tap lever leaks, it will be necessary to renew the tap as a complete unit. It is not possible to dismantle the tap for repair.

4   When reassembling the tap, reverse the procedure for dismantling.

5   Check that the feed pipe from the tap to the carburettor is in good condition and that the push-on joints are a good fit, irrespective of the retaining wire clips. If particles of rubber are found in the filter, replace the pipe, since this is an indication that the internal bore is breaking up.

6   If there have been indications of water contamination in the fuel, the removal of the tap presents a good opportunity to drain and flush the tank completely. Many irritating fuel system faults can be traced to water in the petrol. This often appears as a result of condensation inside the petrol tank. The resulting blobs of water are easily drawn into the carburettor, where they can cause intermittent blockages in the jets and drillings. Any accumulations of water should therefore be flushed from the tank before the tap is refitted. The tubular filter gauze should be removed and cleaned carefully prior to reassembly.

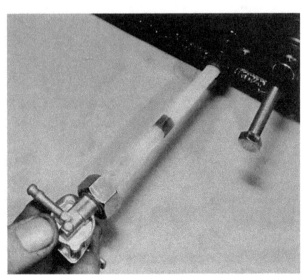

5.6 Gauze filter is attached to petrol tap

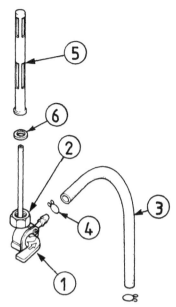

Fig. 2.1 Petrol tap assembly

1	Petrol tap	4	Pipe clip – 2 off
2	Nut	5	Filter
3	Fuel feed pipe	6	Washer

6 Carburettor: removal and refitting

1   Turn the petrol tap to the 'Off' position and free the clips on the petrol pipe, float bowl drain tube, and carburettor air vent tube. Disconnect these three pipes. Slacken and remove the two nuts securing the carburettor to the inlet stub and slacken the clamp securing the air filter hose.

2   Remove the right-hand side panel by unscrewing the front retaining screw and disengaging the panel from its rear mounting on the frame tube. Gently pull the air filter hose back off the carburettor. Slide the carburettor back off its mounting studs and withdraw it from the machine between the battery and the frame lower tube. As soon as the black plastic carburettor top can be reached easily, unscrew the top and carefully remove the throttle valve assembly. Withdraw completely the carburettor. Tape the throttle valve assembly to the frame tube to prevent damage to the throttle valve or needle. If the carburettor is likely to be removed for some time, pack the inlet port with clean rag to prevent the entry of dirt.

3   Refitting the carburettor is a straightforward reversal of the removal sequence. Ensure that the throttle valve is replaced the correct way round, ie with the cutaway facing the air filter, or rear of the carburettor. Do not overtighten the two carburettor/inlet stub mounting nuts or the mating flange will be distorted, thus causing an induction leak. Connect the three pipes and secure their wire clips using a suitable pair of pliers. Turn the petrol tap to the 'On' or 'Res' position as necessary and check carefully for fuel leaks.

4   The tuning procedure for re-setting a carburettor is given in Section 10 of this Chapter. Once this has been carried out, the oil pump cable adjustment must be carefully checked and altered if necessary. It is most important to remember that the oil pump cable adjustment will vary with throttle cable adjustment and must be adjusted to suit.

7 Carburettor: dismantling and reassembly

1   To dismantle the carburettor for examination and cleaning, the carburettor body must first be withdrawn from the machine

as described in Section 6 of this Chapter. Drain any petrol remaining in the float chamber by unscrewing the drain plug at its base. Note that if the presence of dirt or water is suspected in the float chamber it is useful to drain the float chamber into a clean container so that any dirt or water can immediately be seen.

2   Slacken and remove the two float bowl retaining screws and remove the float bowl. It may need a very gentle tap at the joint area to free it. Slide out the float pivot pin and withdraw the float and float needle. The main jet is located at the centre of the carburettor and is identified by its slotted cheese head. It can be unscrewed on its own or together with the hexagon-headed needle jet holder to which it is attached. Both must be removed before the needle jet can be reached. Press the needle jet out from above.

3   The pilot jet is located next to the main jet; on H100 models it is pressed in and should not be disturbed, while on H100 S models it can be unscrewed for cleaning purposes. On H100 models note that the jet must be cleaned while in position, using a blast of compressed air. To remove the throttle stop and pilot air screws, screw in each one until it seats lightly, recording carefully the exact number of turns required to do this, then remove the screw and its spring. Remove the choke (starter) assembly as a single unit; do not attempt to dismantle it further.

4   Turning back to the throttle valve assembly, disconnect the end nipple of the cable from the valve by compressing the return spring with the fingers and sliding the cable down the slot in the side of the valve. Remove the throttle valve, throttle return spring, and carburettor top. If the cable is to be renewed, remove the rubber sealing cap as well.

5   With the throttle valve components set out on a clean sheet of paper on the working surface, remove the needle retaining clip from inside the throttle valve and slide out the needle. Do not disturb the small needle clip unless absolutely necessary as it is easily lost or damaged.

6   Reassembly is a straightforward reversal of the dismantling procedure. Each part must be scrupulously clean and new O-rings must be fitted if required. Screw the pilot air screw in until it seats lightly, then unscrew it the previously noted number of turns. The throttle stop screw, if disturbed, must be fitted in the same way. These settings will return the carburettor to the state of adjustment it was in when stripped, and will serve as a basis for starting the engine and tuning the cleaned and rebuilt carburettor. While working on the carburettor, take great care not to overtighten any of the components as all are delicate and easily damaged.

7.2a Float bowl is retained by two screws

7.2b Push out float pivot pin to permit ...

7.2c ... removal of the float and float needle

7.2d Main jet and needle jet holder can be removed as one piece

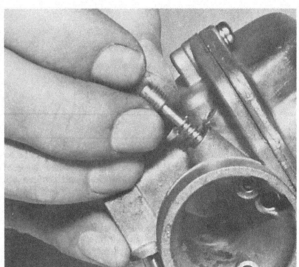

7.3a Record the exact original position of pilot air ...

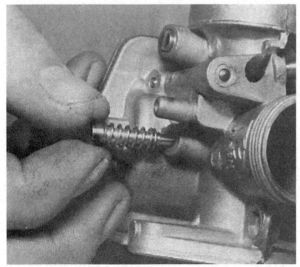

7.3b ... and throttle stop screws before removing them (see text)

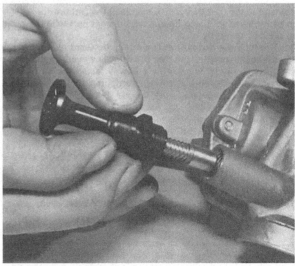

7.3c Choke assembly is removed in one piece

7.5a Throttle valve components

7.5b Note correct position of needle retaining clip

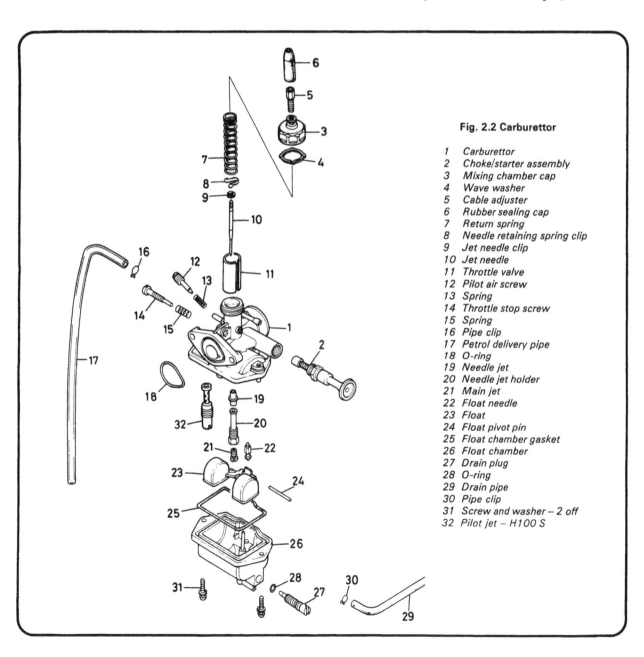

Fig. 2.2 Carburettor

1	Carburettor
2	Choke/starter assembly
3	Mixing chamber cap
4	Wave washer
5	Cable adjuster
6	Rubber sealing cap
7	Return spring
8	Needle retaining spring clip
9	Jet needle clip
10	Jet needle
11	Throttle valve
12	Pilot air screw
13	Spring
14	Throttle stop screw
15	Spring
16	Pipe clip
17	Petrol delivery pipe
18	O-ring
19	Needle jet
20	Needle jet holder
21	Main jet
22	Float needle
23	Float
24	Float pivot pin
25	Float chamber gasket
26	Float chamber
27	Drain plug
28	O-ring
29	Drain pipe
30	Pipe clip
31	Screw and washer – 2 off
32	Pilot jet – H100 S

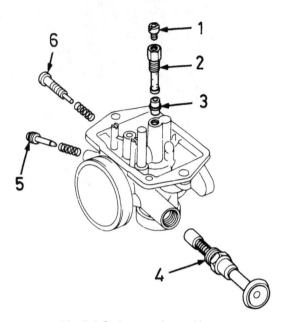

Fig. 2.3 Carburettor jet positions

1   Main jet   4   Choke/starter assembly
2   Needle jet holder   5   Pilot air screw
3   Needle jet   6   Throttle stop screw

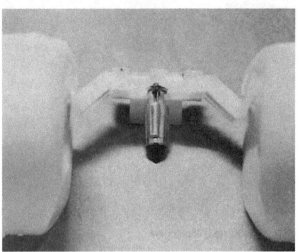

8.1 Float and float needle must be renewed if worn

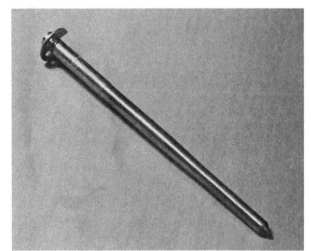

8.5 Throttle needle must be straight and unworn. Check clip position

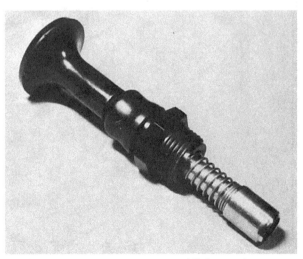

8.6 Choke assembly must be renewed if damaged

8   Carburettor: examination and renovation

1   Having dismantled the carburettor as described in Section 7 the various components should be laid out for examination. If symptoms of flooding have been in evidence, check that the float is not leaking, by shaking and listening for petrol inside. It is rare to find leaks in plastic floats, this problem being more common in the brass type.
2   A more likely cause of flooding is dirt on the float needle or its seat. Examine the faces of the needle and seat for foreign matter and also for scoring. If in bad condition, renew the needle and note whether any improvement is obtained. The valve seat cannot be removed from the body and if badly damaged the entire body must be renewed.
3   The main jet screws into the needle jet, which is central in the carburettor body. It is not prone to any real degree of wear, but can become blocked by contaminants in the petrol. Jets and air passages can be cleared by an air jet, either from an air line or a foot pump. As a last resort, a fine bristle from a nailbrush or similar may be used, but on no account should wire be used as this may damage the precision drilling.
4   The needle jet may become worn after a considerable mileage has been covered and should be renewed along with the needle. Always fit replacement parts as a pair.
5   Examine the throttle valve for scoring or wear, renewing if badly damaged. If damage is evident, check the internal bore of the carburettor, and if necessary renew this also. Check that the needle is free from scoring or other damage and roll it on a flat surface to check that it has not become bent.
6   Examine the choke (starter) plunger assembly for wear. If it is not held firmly in the extended position or if there is damage to the brass plunger or its seating face, the whole assembly must be replaced. No parts are available to recondition this assembly.

9 Carburettor: checking and setting the float height

1   It is important that the level of fuel in the float bowl is maintained at the prescribed height to avoid adverse effects on the mixture strength. It is worth noting that unless the correct float height is set, it will be impossible to set the remaining adjustments to obtain efficient running.
2   The float height is measured between the gasket face of the carburettor body and the bottom of the float. This should be done with the carburettor turned 90° to its normal position so that the weight of the float is not applied to the valve needle. The latter should just bear upon the valve seat when the measurement is made.
3   The correct float height is given in the specifications Section of this Chapter. If the setting is faulty the float and float needle must be removed and examined very carefully for wear. The relevant part or parts must be renewed to produce the correct setting as no adjustment is possible.

10 Carburettor: adjustments and settings

1   The various jet sizes, throttle valve cutaway and needle position are predetermined by the manufacturer and should not require modification. Check with the Specifications list at the beginning of this Chapter if there is any doubt about the types fitted.
2   Before any attempt at adjustment is made, it is important to understand which parts of the instrument control which part of its operating range. A carburettor must be capable of delivering the correct fuel/air ratio for any given engine speed and load. To this end, the throttle valve, or slide as it is often known, controls the volume of air passing through the choke or bore of the instrument. The fuel, on the other hand is regulated by the pilot and main jets, by the jet needle, and to some extent, by the amount of cutaway on the throttle valve.
3   As a rough guide, up to $\frac{1}{8}$ throttle is controlled by the pilot jet, $\frac{1}{8}$ to $\frac{1}{4}$ by the throttle valve cutaway, $\frac{1}{4}$ to $\frac{3}{4}$ throttle by the needle position and from $\frac{3}{4}$ to full throttle by the size of the main jet. These are only approximate divisions, which are by no means clear cut. There is a certain amount of overlap between the various stages.
4   If any particular carburation fault has been noted, it is a good idea to try to establish the most likely cause before dismantling or adjusting takes place. If, for example, the engine runs normally at road speeds, but refuses to tick over evenly, the fault probably lies with the pilot mixture system, and will most likely prove to be an obstructed jet. Whatever the problem may appear to be, it is worth checking that the jets are clear and that all the components are of the correct type. Having checked these points, refit the carburettor and check the settings as follows.
5   Set the pilot air screw to the position given in the Specifications Section. Start the engine, and allow it to attain its normal working temperature. This is best done by riding the machine for a few miles. Set the throttle stop screw to give a normal idling speed. Try turning the pilot mixture screw inwards by about $\frac{1}{4}$ turn at a time, noting its effect on the idling speed, then repeat the process, this time turning the screw outwards. The pilot air screw should be set in the position which gives the fastest consistent tickover. If desired, the tickover speed may be reduced further by lowering the throttle stop screw, but care should be taken that this does not cause the engine to falter and stop after the throttle twistgrip has been opened and closed a few times.
6   Throttle cable adjustment should be checked at regular intervals and after any work is done to the carburettor, oil pump or to the cable itself. Slacken the locknut of the adjuster on the twistgrip and screw the adjuster in to get maximum free play in the cable. The specified throttle cable free play is 2 – 6 mm ($\frac{1}{8}$

– $\frac{1}{4}$ in) measured at the inner flange of the twistgrip rubber. To measure this, use a piece of chalk or some paint to mark both the twistgrip rubber at its inner flange, and the twistgrip drum. These two marks will provide a convenient reference point for future adjustment. Carefully open the throttle by rotating the twistgrip rubber in the usual way until all the free play in the cable has been taken up. Measure the distance around the circumference of the drum between the static mark on the drum and the mark that has moved with the twistgrip rubber. If this distance is more or less than the specified amount, use the adjuster on the carburettor top to adjust the cable as necessary. The adjuster on the twistgrip may be used if necessary to complete the operation, but note that it is normally only used for minor adjustments. Tighten the adjuster locknuts, slide the rubber sealing sleeves back over the adjusters and fully open and close the throttle several times. Check that the cable adjustment has remained the same. Remember that oil pump cable adjustment will be altered whenever the throttle cable is adjusted, and that the oil pump therefore, must be checked at the same time as described in Section 18 of this Chapter.

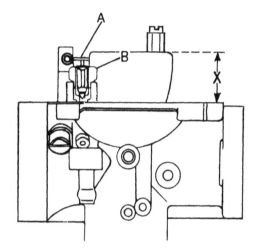

Fig. 2.4 Checking the float height

A   Float tongue
B   Float valve
X   13.5 mm (0.53 in)

10.5a Location of pilot air screw (arrowed)

10.5b Location of throttle stop screw

11.3 Reed valve assembly must be renewed if damaged or worn

11 Reed valve asembly: removal, examination and refitting

1   Remove the oil feed pipe from the inlet stub union and plug it to prevent the loss of oil or the entry of air or dirt. Remove the carburettor as described in Section 6 of this Chapter. Slacken and remove the four inlet stub retaining bolts and displace the stub. It may need a gentle tap using a soft-faced mallet to free it. Remove the reed valve assembly. Take great care when handling the reed valve assembly as it is easily damaged.

2   The reed valve assembly provides a supplementary method of controlling the inlet timing which functions in addition to the normal piston porting arrangement. The normal piston-ported intake tract is designed to open earlier and close later than is usual in engines of this type, thus producing much better high-speed performance at the expense of the low to medium speed ranges. The reed valve operates automatically, opening and closing as a result of the combination of atmospheric pressure and piston position, which in practice has the effect of making the engine more efficient at the low to medium speed ranges. Piston porting and reed valve therefore combine to give the engine more power at all engine speeds than can be available to a conventionally-ported unit. The reed valve has an added advantage in that it eliminates the blow-back of air-fuel mixture through the carburettor which is an inevitable product of a piston-ported engine, thus reducing petrol consumption.

3   As far as maintenance is concerned, the reed valve requires none as it is extremely simple in construction and is automatic in operation. However it is also extremely delicate and must be kept clean at all times and handled very carefully. Check the whole assembly for cracks or other signs of wear, and make sure that the reed petals seat firmly on the rubber valve seat. Any sign of damage at all will mean the whole assembly will have to be renewed. No repairs are possible, and it is not advisable to attempt to strip the assembly further as no individual components are available. Furthermore do not attempt to modify the assembly by bending the stopper plate or by any other means as this will at least adversely affect engine performance. More probably the stress limits will cause them to crack, allowing the pieces to drop straight into the engine with disastrous consequences.

4   Refitting the reed valve assembly is a straightforward reversal of the removal procedure. Always fit new gaskets to prevent induction leaks and be careful not to overtighten the inlet stub bolts. When connecting the oil feed pipe, if the end of

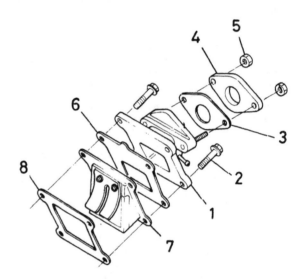

Fig. 2.5 Reed valve assembly

1 Inlet stub	5 Nut – 2 off
2 Bolt – 4 off	6 Gasket
3 Gasket	7 Reed valve block
4 Spacer	8 Gasket

the pipe is full of oil when it is unplugged, bleeding air from the system will not be necessary. Simply push the end of the pipe over the inlet stub union and secure it with the wire clip. If however, no oil is visible, the feed pipe will have to be bled as described in Section 19 of this Chapter.

12 Air filter: removal, examination and refitting

1   The air used in combustion is drawn into the carburettor via an air filter element. This performs the vital task of removing dust and any other airborne impurities which would otherwise enter the engine, causing premature wear. It follows that the

element must be kept clean and renewed if damaged, as it will have an adverse effect on performance if neglected. Apart from the obvious problem of increased wear caused by a damaged element, a clogged or broken filter will upset the mixture setting, allowing it to become too rich or too weak.

2   To gain access to the filter element slacken and remove the three screws securing the left-hand sidepanel and withdraw the side panel. This exposes the metal frame which supports the filter element and is retained by a single screw. Slacken and

remove the screw, then withdraw the frame and the element itself.

3   The element should be cleaned by washing it in a high flash point solvent such as white spirit. Squeeze the element dry, but do not wring it out as this will damage the foam. Soak the cleaned, dry, element in clean gear oil (SAE 80 or 90) and squeeze the surplus out. The element must be only slightly oily to the touch to be properly effective. The air cleaner assembly can then be refitted by reversing the removal procedure.

12.2a Air filter cover is retained by three screws

12.2b Element is supported by a metal frame secured by a single screw

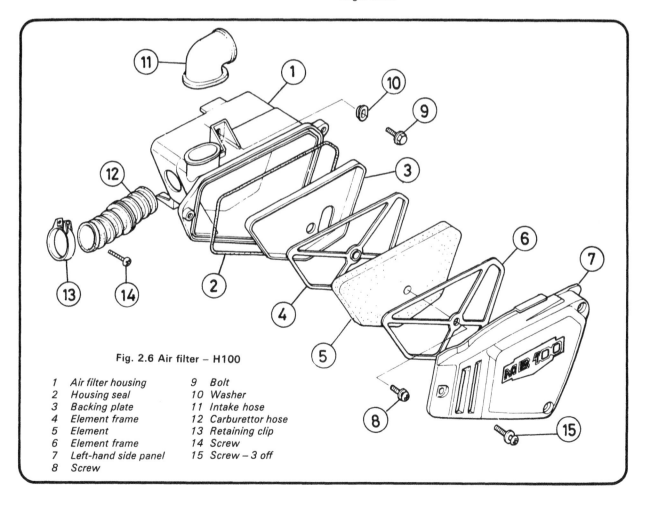

Fig. 2.6 Air filter – H100

1	Air filter housing	9	Bolt
2	Housing seal	10	Washer
3	Backing plate	11	Intake hose
4	Element frame	12	Carburettor hose
5	Element	13	Retaining clip
6	Element frame	14	Screw
7	Left-hand side panel	15	Screw – 3 off
8	Screw		

12.2c The element must be regularly cleaned and re-oiled

13 Exhaust system: removal, examination and refitting

1   On H100 models the exhaust system consists of a one-piece unit comprising exhaust pipe and silencer, while on H100 S models the two are separate, being joined by a clamp and tubular gasket. A removable baffle tube is fitted in the silencer tailpipe for cleaning purposes.
2   To remove the system, unscrew the single bolt which secures the front of the silencer to the footrest bracket (H100 S models only), remove the two nuts securing the pipe to the cylinder barrel, then remove the single rear mounting bolt and withdraw the system as a single unit. On H100 S models the clamp is then slackened so that the pipe and silencer can be separated; on reassembly, always renew the gasket and do not fasten the clamp until all mountings have been tightened securely. Remove its retaining screw and spacer, and withdraw the baffle tube.
3   Baffle tubes are well known for their tendency to stick in the silencer as carbon or rust deposits build up around them and jam them in place. If such a case is found, try rotating the baffle to free it. If the tube remains stuck in position some means will have to be found of removing it without causing too much damage to the tube or to the silencer. If necessary take the assembly to a dealer for expert advice.
4   Once removed, carefully check the exhaust system for carbon build-up. This will occur mainly in the immediate area of the exhaust port and at the other end in the baffle tube and tailpipe. Remove all traces of this with a scraper or with a rotary wire brush attachment fitted to an electric drill. The component most likely to cause trouble and therefore require attention is the silencer baffle, which will block up with a sludge composed of carbon and oil if not cleaned out at regular intervals. A two-stroke engine is very susceptible to this fault.
5   If the build-up of carbon and oil is not too great, a wash with a petrol/paraffin mix will probably suffice as a cleaning medium. Otherwise more drastic action will be necessary, such as the application of a blow lamp to burn away the accumulated deposits. Before the baffle is refitted it must be completely clean with none of the holes in the baffle obstructed.
6   When replacing the baffle, make sure that the retaining bolt is located correctly and fully tightened. If the baffle bolt falls out, the baffle will work loose, creating excessive exhaust noise accompanied by a marked fall-off in performance.
7   Do not run the machine without the baffle in the silencer or modify the baffle in any way. Although the changed exhaust note may give the illusion of increased power, the chances are that the performance will be reduced, accompanied by a noticeable lack of acceleration. This is also a risk of prosecution by causing an excessive noise. The carburettor is jetted to take into account the fitting of a silencer of a certain design and if this balance is disturbed, the carburation will suffer accordingly.
8   Refitting is a straightforward reversal of the removal procedure. Always renew the exhaust port gasket to prevent leaks and tighten the nuts at the front flange first to ensure that the exhaust system is positioned correctly.

Fig. 2.7 Exhaust system – H100

1 Exhaust system	3 Sealing ring	5 Bolt	7 Screw
2 Baffle tube	4 Nut – 2 off	6 Spacer	

13.2a Exhaust system is secured by two nuts at the front flange ...

13.2b ... and a single bolt at the rear mounting

13.2c Baffle is retained by a single screw

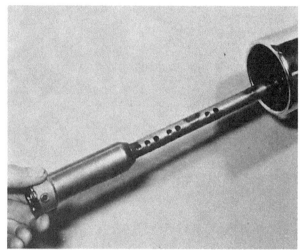

13.2d All traces of carbon should be removed from the baffle tube

13.2e Exhaust pipe/silencer joint is secured by a clamp – H100 S

14 Oil injection system: description

The oil injection system employed by Honda on this model is a simple one, designed to avoid the inconvenience of having to mix petrol and oil in the petrol tank for engine lubrication. The system consists of a separate oil-carrying reservoir which has its own filler cap and outlet, and which is provided with an oil level gauge unit or sight glass to give an instant reading of the amount of oil remaining in the tank. Oil is fed from this reservoir to a mechanical pump situated on the crankcase and driven from the engine by reduction gear. The pump delivers oil at a predetermined rate via a synthetic rubber feed pipe to an oilway in the inlet stub. In consequence the oil is carried into the engine by the incoming charge of air/fuel mixture from the carburettor. The pump's output is varied according to the throttle position by a control cable linked to the throttle cable at a junction box; this cable operating a control lever on the pump body.

In this way the big-end bearing, left-hand main bearing, small-end bearing, piston and cylinder bore always receive the correct amount of oil according to their needs, a much more efficient system of lubrication which shows itself in a less smoky exhaust and much longer intervals between de-carbonising operations than is possible for a petroil-lubricated two-stroke engine unit.

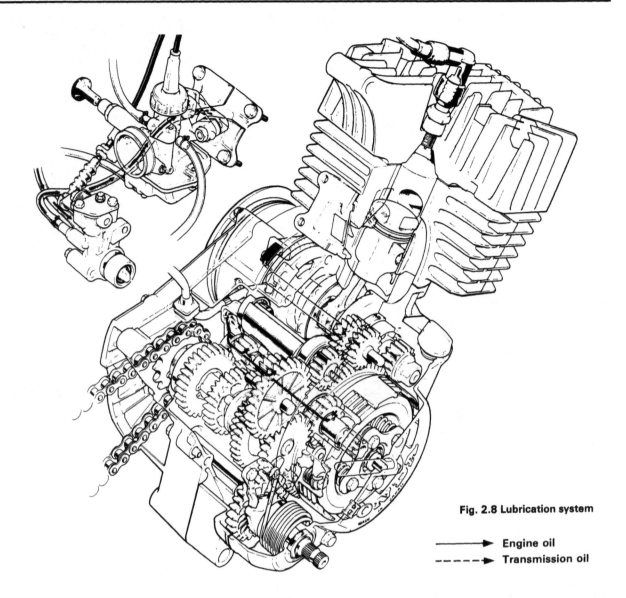

Fig. 2.8 Lubrication system

———▶   Engine oil
– – – –▶   Transmission oil

15 Oil filter: removal, examination and cleaning

1   Before the oil filter can be removed it is first necessary to drain the tank. This is easily accomplished by disconnecting the oil tank/oil pump feed line at the outlet first beneath the tank and allowing the oil to drain into a clean container. Alternatively the tank can be removed from the machine, as described in Section 2 of this Chapter, in order to give better access to the components concerned.

2   Once the tank is fully drained, slacken the clip at the base of the tank and withdraw the oil filter assembly. The filter gauze itself can then be detached and cleaned by blowing from the inside with compressed air, or flushing in clean petrol (gasoline).

3   If a large amount of dirt is found in the filter, the tank should be cleaned as described in Section 3 of this Chapter. Although this can be done with the tank in situ, the task is much easier if the tank is removed first.

4   Refitting the oil filter is a straightforward reversal of the removal procedure. Refill the tank with clean oil and carry out the bleeding procedure to expel air from the oil tank/oil pump feed line and from the oil pump itself. Once the bleeding procedure is complete, securely fasten all clips and ensure that there are no oil leaks.

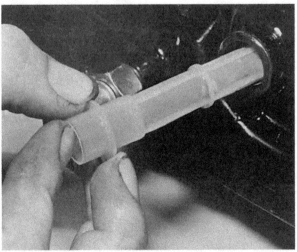

15.2 Gauze oil filter is located in the oil tank outlet

16 Oil pump: removal and refitting

1   To remove the oil pump, first disconnect the oil tank/oil pump feed line at the outlet first beneath the tank by using a suitable pair of pliers to release the wire clip and by pulling the pipe gently away. Place a finger over the tank outlet, temporarily to stop the flow of oil. Plug the ends of the two pipes using screws or bolts of suitable size, taking great care not to damage the pipes.

2   Disconnect the oil pump cable by slackening the two nuts at the adjuster bracket and sliding the cable clear of the bracket, whereupon the cable end nipple can be disengaged from the oil pump control lever. Disconnect the oil pump/inlet stub feed pipe at the stub union and plug the pipe using a screw or bolt of suitable size. Slacken and remove the two oil pump retaining bolts and withdraw the pump.

3   Refitting is a straightforward reversal of the removal procedure. Examine the O-ring on the oil pump mounting spigot and renew it if it is worn or damaged. Apply a little grease to the pump mounting spigot and align the blade on the pump with the slot in the drive gear shaft. Ensure that the pump mounting

bolts are tightened securely and connect the oil tank/oil pump feed line after unplugging it. Ensure that the oil pump/inlet stub feed line is secured to the pump by its wire clip and route the feed line through the clamp on the adjuster bracket. Leave the end free at the inlet stub union to enable bleeding to be carried out. Refit the oil pump cable and adjust it as described in Section 18 of this Chapter. The system must now be cleared of air by bleeding as described in Section 19.

17 Oil pump: examination

1   When the oil pump has been removed as described in Section 16 of this Chapter, carefully examine it for damage to the main body casting and for correct operation of the control lever. Check that the O-ring around its fitting spigot is in good condition. Replace it if necessary.

2   If there are any signs of damage, or if you have good reason to suspect that the pump is not working properly, discard it and fit a new one. Stripping the pump for repair is not a practical proposition as there are no parts available to recondition it.

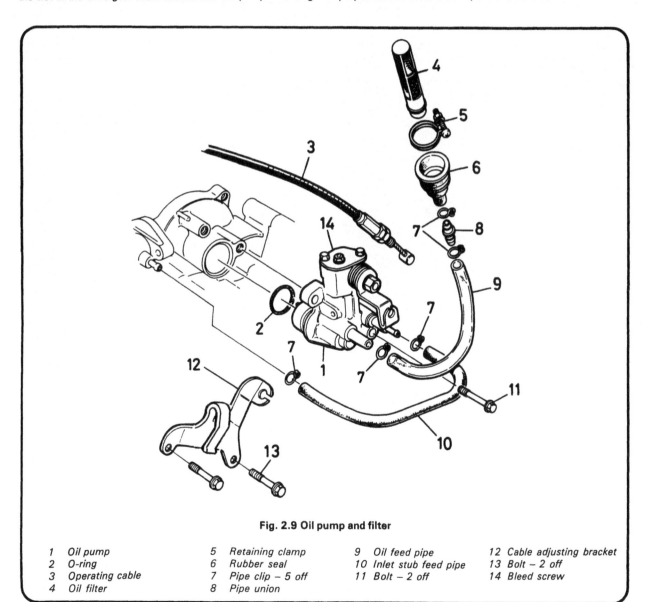

Fig. 2.9 Oil pump and filter

1 Oil pump	5 Retaining clamp	9 Oil feed pipe	12 Cable adjusting bracket
2 O-ring	6 Rubber seal	10 Inlet stub feed pipe	13 Bolt – 2 off
3 Operating cable	7 Pipe clip – 5 off	11 Bolt – 2 off	14 Bleed screw
4 Oil filter	8 Pipe union		

18 Oil pump: cable adjustment

1   The oil pump cable adjustment must be checked at the specified service intervals and whenever the carburettor, oil pump or throttle cable are disturbed. First the throttle cable must be adjusted correctly as described in Section 10 of this Chapter. Open the throttle fully and check that the reference mark on the pump control lever lines up exactly with the index mark on the pump body. Note that there is another line situated approximately half-way between the lever fully closed stop and the full open reference mark. This is a factory setting mark and is to be ignored at all times. Use only the fully open reference mark when adjusting the oil pump cable. Refer to the accompanying photographs to ensure that the correct mark is used.

2   If the two marks do not align exactly with the throttle fully open, slacken the adjuster locknut and turn the adjuster nut as required to bring the marks into line. Once the cable is adjusted correctly, tighten the adjuster locknut and fully open and close the throttle two or three times to check the lever operation and to settle the cable. Check that the adjustment has remained the same.

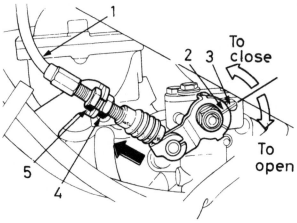

Fig. 2.10 Oil pump cable adjustment

1	Operating cable	4	Locknut
2	Control lever reference mark	5	Adjusting nut
3	Index mark		

18.1 Oil pump setting marks (arrowed) must align exactly when throttle is in fully open position

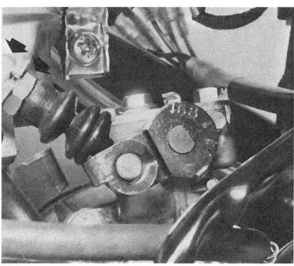

18.2 Use adjusting nuts to bring marks into line if necessary

19 Oil injection system: the bleeding procedure

1   Bleeding must be carried out whenever any part of the oil injection system is disturbed, if the oil in the tank has been allowed to drain completely or run dangerously low, or if you have any reason to suspect the presence of air in the system. Air in the system will rapidly produce an airlock which will interrupt the constant supply of oil, resulting in severe engine damage due to the consequent loss of lubrication. The oil tank must be kept full of oil at all times during this operation.

2   The bleeding operation consists of two parts, removing air from the oil tank, the oil tank/oil pump feed line, and from the pump itself, and removing air from the delivery side of the pump and from the oil pump/inlet stub feed line.

3   To complete the first part, thoroughly clean the oil pump and the area of crankcase around it and pack clean rag around the base of the pump. Check that the oil tank is full of oil, topping it up to the base of the filler neck if necessary. Slacken and remove the hexagon-headed bleed screw which is situated almost centrally in the flat plate on the top of the pump. Oil will flow from the orifice thus exposed. Watch carefully until you can see no more air bubbles in the oil, check that no more bubbles are visible in the transparent feed line, and then replace and tighten down the bleed screw. Remove the rag and clean away the surplus oil.

4   The first part of the bleeding operation can be speeded up if the oil tank/oil pump feed line has been completely drained of oil or is full of air bubbles. Disconnect the feed line at the lower, pump body, union and allow the tank to drain into a clean container until the pipe is full of oil and all air bubbles have been expelled. This will fill the pipe much faster and flush out any air bubbles, if present, in the base of the tank, the oil filter assembly and the oil feed pipe itself. Once the feed line is full of oil, place a finger over the end temporarily to stop the flow of oil, reconnect it to the union on the pump body and secure the wire clip. Pour the drained oil, if totally clean, back into the tank to

avoid wasting it and proceed to clear the pump of air as described in the previous paragraph. If this method is used, check that no air bubbles subsequently emerge in the feed line.

5   Once the oil tank, oil tank/oil pump feed line, and the oil pump itself have been cleared of air bubbles, proceed with the second part of the bleeding operation. This involves running the engine because the only way to expel air from the delivery side of the pump and from the oil pump/inlet stub feed line is to pump oil through, running the engine to operate the pump. The engine will therefore be running for a while with its normal supply of lubricant disconnected and an alternative method of lubrication must be used if the engine is not to suffer serious damage due to lack of lubrication.

6   Place the machine on its centre stand on level ground and in a well-ventilated area. This last is essential as excessive quantities of exhaust smoke and fumes will be produced. Check the oil level in the tank and top up to the base of the filler neck if necessary. Disconnect the petrol pipe at the tap and drain any petrol remaining in the petrol tank into a clean container. Mix up a small quantity (about a pint)of 25/50:1 petroil mixture and pour this into the tank. Connect the petrol pipe up again and turn the petrol tap to the 'Res' position. This will provide the alternative method of lubrication referred to in the previous paragraph, lubricating the engine components as the mixture is consumed during combustion. Disconnect the oil pump/inlet stub feed line at the inlet stub union if this has not already been done. Start the engine and slowly warm it up until it will tick over smoothly. Using one finger, move the oil pump control lever around to the fully open position and keep it there. This will ensure that the pump is working at maximum capacity at idle speed and will hasten the process as much as is advisable. Keep the engine at as low a speed as possible, preferably at a constant idle, to minimise the risk of damage due to overheating while running at a standstill. Watch the end of the oil feed pipe very carefully. If it has been completely emptied, it may take as long as 10 minutes for the oil to appear.

7   When oil does appear in the end of the pipe, continue the bleeding process until you are certain that no more air bubbles are appearing and that all traces of air have therefore been expelled. Due to the slow rate of delivery of the oil pump (less than 0.10 cc/min at idle speed) this operation is very time-consuming and dull. It might help to ease the boredom by wedging open the oil pump control lever with a piece of wood; if this is done carefully the control lever will remain in the fully open position, leaving your hands free for other tasks. The operation must, however, be carried out very carefully if the engine is not to suffer severe and premature damage due to the lack of lubrication which would result were air permitted to remain in the injection system.

8   As soon as the oil flowing from the end of the feed pipe is completely free of air bubbles, reconnect the oil feed pipe to the inlet stub union, secure its wire clip, and stop the engine. Top up the oil tank to just below the filler neck and replace the filler cap. Ensure that all oil pipe unions are plugged firmly into place and that they are secured by their wire clips. Check that the oil lines are routed correctly and secured by such clamps or cable ties are provided for this purpose. Wash off any surplus oil and check that the rag has been removed from the base of the oil pump. Drain any remaining petroil mixture from the tank and replace it with the fresh petrol originally taken out. Connect the petrol pipe up and secure it with the wire clip. Check that the oil pump control lever has returned to the fully closed position and then fully open and close the throttle several times to settle the cable and to check that the pump control lever is operating correctly. Recheck the oil pump cable adjustment as described in Section 18.

9   When taking the machine out on the road after bleeding the oil injection system, remember that the exhaust will be excessively smoky until the surplus oil has been used or burned up. Remember also to check for oil leaks in the system, these being readily apparent if the surplus oil was washed off as described. Any such leaks should be corrected immediately, before the machine is used further.

19.3 Removing bleed screw to bleed oil tank/oil pump feed line

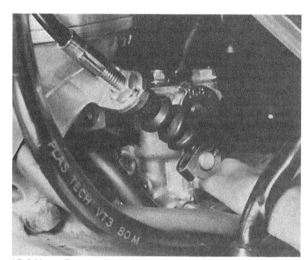

19.6 Use a finger to keep the pump control lever in the fully open position for bleeding

20 Gearbox lubrication: description and maintenance

1   The gearbox oil is contained inside the crankcases and lubricates the crankshaft right-hand main bearing, the clutch and primary drive, and the gearbox and ancillary components.

2   The crankcase reservoir is reached through a screwed plug in the right-hand outer cover and drained through a single drain plug situated in the underside of the crankcase. The level is checked by removing a small level plug in the right-hand outer cover immediately in front of the kickstart shaft.

3   Maintenance is restricted to periodic checks of the oil level and to regular changing of the oil itself. The machine must always be standing upright on level ground whenever work is undertaken on the gearbox oil; it is only in this position that the maximum amount of oil can be removed by draining, and it is also only in this position that the level can be checked accurately. Routine oil changes are made more effective if the machine is fully warmed up to normal operating temperature before the drain plug is removed. The oil is much thinner at high temperatures, enabling more rapid and more complete draining to take place, and any particles of dirt and swarf are held in suspension in the oil and are therefore much more likely to be removed.

4   Once the drain plug has been removed and the oil drained as completely as possible, check the condition of the drain plug sealing washer, renewing it if necessary, and refit the drain plug. Tighten it down to a torque setting of 2.0 – 2.5 kgf m (14 – 18 lbf ft). On refilling, note that different amounts are given for engine reassembly and for routine oil changes. The difference is the amount of residual oil left in the crankcases after routine draining. Temporarily replace the level and filler plugs, start the engine and warm it up to normal operating temperature. With the machine fully warmed up and placed on its centre stand on level ground, stop the engine and remove the level plug after leaving the machine for a few minutes to allow the oil to settle. Oil should merely trickle from the level plug orifice. If no oil appears, add a small quantity through the filler plug orifice until it reaches the required level. If too much oil flows out, allow it to carry on draining through the level plug orifice until the flow is reduced to a gentle trickle.

5   Once the oil level is known to be correct, replace the level and filler plugs and check that all three gearbox plugs are securely tightened. Remove any surplus oil and subsequently check that no oil leaks appear.

20.2 Gearbox drain plug is situated centrally in the underside of the crankcase

20.4a Oil should trickle slowly from the level plug orifice if level is correct

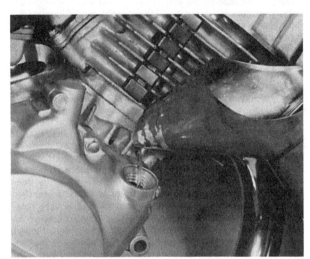

20.4b Top up with recommended grade of oil if necessary

Chapter 3 Ignition system

For specifications and information relating to the H100 S II model, refer to Chapter 7

Contents

Specifications

Ignition system

Type:

H100	Capacitor discharge ignition (CDI)
H100 S	Contact breaker

Ignition timing ....

$15° \pm 3°$ BTDC (@ 3000 rpm – H100)

Ignition HT coil

Minimum spark gap	6 mm (0.24 in)
Winding resistance:	
Primary	0.2 – 0.3 ohm
Secondary	3.4 – 4.2 K ohm

Condenser capacity – H100 S ....

0.27 – 0.33 microfarad

Contact breaker gap – H100 S

Standard	0.35 mm (0.014 in)
Tolerance	0.30 – 0.40 mm (0.012 – 0.016 in)

Spark plug

	NGK	ND
Make		
Type:		
Hot	BR6HS	W20FSR
Standard	BR7HS	W22FSR
Cold	BR8HS	W24FSR
Gap	0.6 – 0.7 mm (0.024 – 0.028 in)	

1 General description

The Honda H100A is equipped with a CDI (capacitor discharge ignition) system. The system is powered by a source coil built into the generator stator. Power from this coil is fed directly to the CDI unit mounted beneath the frame, where it passes through a diode which converts it to direct current (dc). The charge is stored in a capacitor at this stage.

The spark is triggered by the pulser assembly which is mounted on the stator plate. As the magnetic rotor passes the pulser coil a small alternating current (ac) pulse is induced. This enters the CDI unit where it is rectified by a second diode. The heart of the CDI unit is a component known as a thyristor. It acts as an electronic switch, which remains off until a small current is applied to its gate terminal. This causes the thyristor to become conductive, and it will remain in this state until any stored charge has discharged through the primary windings of the coil.

The sudden discharge of low-tension energy through the coil's primary windings in turn induces a high tension charge in the secondary coil. It is this which is applied to the centre electrode of the sparking plug, where it jumps the air gap to earth, igniting the fuel/air mixture.

As engine speed rises, it becomes necessary for the timing of the ignition spark to be advanced in relation to the crankshaft to allow sufficient time for the combustion of the air/fuel mixture to take place at the optimum position. This function is controlled by the voltage build-up time in the pulser, which varies with the speed of the engine, acting on the thyristor gate circuit.

Unusually, the H100 S has reverted to the much simpler contact breaker – triggered type of ignition system. The contact breaker is controlled by a cam on the generator rotor and power is supplied by a source coil on the generator stator, a condenser being fitted in the low-tension circuit to prevent arcing across the points when they separate. Apart from the ignition HT coil, ignition switch and wiring, the only other components in the system are two diodes which prevent current flow reversal between the ignition switch, HT coil and generator components and between the ignition system and other circuits.

The bulk of this Chapter is devoted to the more sophisticated procedures necessary to trace and eliminate faults in the H100 CDI system; refer to the Fault Diagnosis section at the front of this Manual when tracing a fault in the H100 S contact breaker system. Details of maintenance, including contact breaker checking and renewal and ignition timing checking and resetting, are given in the relevant Section of Routine maintenance.

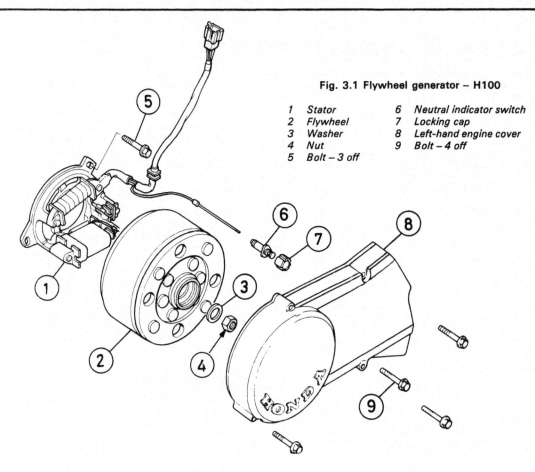

Fig. 3.1 Flywheel generator – H100

1	Stator	6	Neutral indicator switch
2	Flywheel	7	Locking cap
3	Washer	8	Left-hand engine cover
4	Nut	9	Bolt – 4 off
5	Bolt – 3 off		

2 CDI system: fault diagnosis

1   As no means of adjustment is available, any failure of the system can be traced to the failure of a system component or a simple wiring fault. Of the two possibilities, the latter is by far the most likely. In the event of failure, check the system in a logical fashion, as described below.

2   Remove the sparking plug, giving it a quick visual check noting any obvious signs of flooding or oiling. Fit the plug into the plug cap and rest it on the cylinder head so that the metal body of the plug is in good contact with the cylinder head metal. The electrode end of the plug should be positioned so that sparking can be checked as the engine is spun over using the kickstart.

3   *Important note.* The energy levels in electronic systems can be very high. On no account should the ignition be switched on whilst the plug or plug cap is being held. Shocks from the HT circuit can be most unpleasant. Secondly, it is vital that the plug is in position and soundly earthed when the system is checked for sparking. The CDI unit can be seriously damaged if the HT circuit becomes isolated.

4   Having observed the above precautions, turn the ignition switch to 'On' and kick the engine over. If the system is in good condition a regular, fat blue spark should be evident at the plug electrodes. If the spark appears thin or yellowish, or is non-existent, further investigation will be necessary. Before proceeding further, turn the ignition off and remove the key as a safety measure.

5   Ignition faults can be divided into two categories, namely those where the ignition system has failed completely, and those which are due to a partial failure. The likely faults are listed below, starting with the most probable source of failure. Work through the list systematically, referring to the subsequent sections for full details of the necessary checks and tests.

Total or partial ignition system failure
  a)   Loose, corroded or damaged wiring connections, broken or shorted wiring between any of the component parts of the ignition system
  b)   Faulty main switch
  c)   Faulty ignition coil
  d)   Faulty CDI unit
  e)   Faulty generator

3 CDI system: checking the wiring

1   The wiring should be checked visually, noting any signs of corrosion around the various terminals and connectors. If the fault has developed in wet conditions it follows that water may have entered any of the connectors or switches, causing a short circuit. A temporary cure can be effected by spraying the relevant area with one of the proprietary de-watering aerosols, such as WD40 or similar. A more permanent solution is to dismantle the switch or connector and coat the exposed parts with silicone grease to prevent the ingress of water. The exposed backs of connectors can be sealed off using a silicone rubber sealant.

2   Light corrosion can normally be cured by scraping or sanding the affected area, through in serious cases it may prove necessary to renew the switch or connector affected. Check the wiring for chafing or breakage, particularly where it passes close to part of the frame or its fittings. As a temporary measure, damaged insulation can be repaired with PVC tape, but the wire concerned should be renewed at the earliest opportunity.

3   Using the wiring diagram at the end of the manual, check each wire for breakage or short circuits using a multimeter set on the resistance scale or a dry battery and bulb wired as shown in the accompanying illustration. In each case, there should be continuity between the ends of each wire.

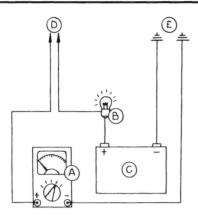

Fig. 3.2 Method of checking the wiring

A   Multimeter   D   Positive probe
B   Bulb   E   Negative probe
C   Battery

4   CDI system: checking the ignition switch

1   The ignition system is controlled by the ignition switch or main switch which is retained in the instrument console by plastic clips. The switch has four terminals and leads, of which two are involved in controlling the ignition system. These are the 'IG' terminal (black/white lead) and the 'E' terminal (green lead). The two terminals are connected when the switch is in the 'Off' position and prevent the ignition system from functioning by shorting the CDI unit to earth. When the switch is in the 'On' position the CDI/earth connection is broken and the system is allowed to function.

2   If the operation of the switch is suspect, reference should be made to the wiring diagram at the end of this book. The switch connections are also shown in diagrammatic form and indicate which terminals are connected in the various switch positions. The wiring from the switch can be traced back to the respective connectors where test connections can be made most conveniently.

3   The purpose of the test is to check whether the switch connections are being made and broken as indicated by the diagram. In the interests of safety the test must be made with the machine's battery disconnected, thus avoiding accidental damage to the CDI system or the owner. The test can be made

with a multimeter set on the resistance scale, or with a simple dry battery and bulb arrangement as previously shown. Connect one probe lead to each terminal and note the reading or bulb indication in each switch position.

4   If the test indicates that the black/white lead is earthed irrespective of the switch position, trace and disconnect the ignition (black/white) and earth (green) leads from the ignition switch. Repeat the test with the switch isolated. If no charge is apparent, the switch should be considered faulty and renewed. If the switch works normally when isolated, the fault must lie in the black/white lead between the switch and the CDI unit.

5   The ignition switch is removed by unclipping it from the instrument panel/speedometer asssembly, having first removed the headlamp assembly and instrument panel as described in the relevant Section of Chapter 4. While it is a sealed unit and can only, officially, be repaired by renewing it as a complete assembly; there is nothing to be lost by attempting to repair it if tests have proven it faulty. Depending on the owner's skill, worn contacts may be reclaimed by building up with solder or in some cases, merely cleaning with WD40 or a similar water dispersant spray.

5   Ignition coil: location and testing

1   The ignition coil is a sealed unit, and will normally give long service without need for attention. It is mounted beneath the frame gusseting to the rear of the steering head, or on the frame top tube, and is covered in use by the fuel tank. It follows that it will be necessary to remove the tank in order to gain access to the coil.

2   If a weak spark and difficult starting causes the performance of the coil to be suspect, it should, in general, be tested by a Honda service agent or an auto-electrical expert. They will have the necessary appropriate test equipment. It is, however, possible to perform a number of basic tests, using a multimeter with ohms and kilo ohms scales. The primary winding resistance should be checked by connecting one of the meter probe leads to the spade terminal and the other earthed against the coil mounting lug. The secondary windings are checked by connecting the probe leads to the high tension lead, having removed the plug cap, and to the coil mounting lug. Check the readings thus obtained against the figures given in the specification Section of this Chapter.

3   Should any of these checks not produce the expected result, the coil should then be taken to a Honda service agent or auto-electrician for a more thorough check. If the coil is found to be faulty, it must be replaced; it is not possible to effect a satisfactory repair.

4.5 Ignition switch is a sealed unit mounted in the instrument panel

5.1 Location of ignition components – ignition coil and CDI unit – H100

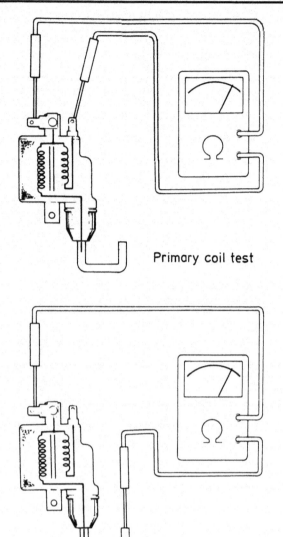

Primary coil test

Secondary coil test

Fig. 3.3 Ignition coil resistance test

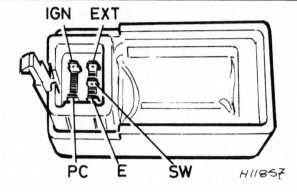

IGN EXT

PC   E   SW   H11857

TESTER⊕ / TESTER⊖	SW	EXT	PC	E	IGN
SW		∞	∞	∞	∞
EXT	0.1 – 10		∞	∞	∞
PC	0.5 – 200	0.5 – 50		1.0 – 50	∞
E	0.2 – 30	0.1 – 10	∞		∞
IGN	∞	∞	∞	∞	

Measuring ranges: SANWA: x 1 k Ω
                   KOWA: x 100 Ω

Fig. 3.4 CDI unit testing

The diagram illustrates the CDI unit connections referred to in the table. For owners not possessing a test meter the unit or the complete machine can be taken to a Honda Service Agent for testing.

4 If the CDI unit is found to be faulty as a result of these tests, it must be renewed. No repairs are possible. It might be worthwhile trying to obtain a good secondhand unit from a motorcycle breaker, in view of the cost of a new part.

7 Flywheel generator: testing

1 The ignition system is powered by a source coil and triggered by a pulser coil, both of which are built into the generator stator plate mounted on the left-hand crankcase. It follows that if the generator malfunctions, it will affect the operation of the ignition system, possibly without affecting the remainder of the electrical system. Six leads exit from the stator assembly. Of these identify the plain green, the black/red, and the blue/yellow wires.

2 Disconnect the wires at their connectors. Using a multimeter set on the ohms scale measure the resistances between the terminals of those wires. The values should be as shown below.

Black/red to green lead (source coil)   50 – 300 ohm
Blue/yellow to green lead (pulser coil) 10 – 100 ohm

If the readings obtained fall outside the limits given the generator stator is faulty. Unfortunately this means that the complete flywheel generator must be renewed; individual generator components are not available as spare parts. A check should be made to ensure that the fault is not due to a broken or damaged wire which could be repaired easily. If no apparent fault is found, it must be assumed that the coil or coils which gave the wrong reading has indeed malfunctioned. In such a case it is advised that the generator assembly be taken to an authorised Honda dealer for confirmation of this before a new assembly is purchased, as this is likely to be very expensive. Once more, a search around the motorcycle breakers might reveal a good secondhand assembly.

6 CDI unit: location and testing

1 The CDI unit takes the form of a sealed metal box mounted beneath the fuel tank. In the event of malfunction the unit may be tested in situ after the fuel tank has been removed and the wiring connectors traced and separated. Honda advise against the use of any test meter other than the Sanwa Electric Tester (Honda part number 07308-0020000) or the Kowa Electric Tester (TH-5H), because they feel that the use of other devices may result in inaccurate readings.

2 Most owners will find that they either do not possess a multimeter, in which case they will probably prefer to have the unit checked by a Honda Service Agent, or own a meter which is not of the specified make or model. In the latter case, a good indication of the unit's condition can be gleaned in spite of inaccuracies in the readings. If necessary, the CDI unit can be taken to a Honda Service Agent or auto-electrical specialist for confirmation of its condition.

3 The test details are given in the accompanying illustration in the form of a table of meter probe connections with the expected reading in each instance. If an ordinary multimeter is used the resistance range may be determined by trial and error.

Electrode gap check – use a wire type gauge for best results

Electrode gap adjustment – bend the side electrode using the correct tool

Normal condition – A brown, tan or grey firing end indicates that the engine is in good condition and that the plug type is correct

Ash deposits – Light brown deposits encrusted on the electrodes and insulator, leading to misfire and hesitation. Caused by excessive amounts of oil in the combustion chamber or poor quality fuel/oil

Carbon fouling – Dry, black sooty deposits leading to misfire and weak spark. Caused by an over-rich fuel/air mixture, faulty choke operation or blocked air filter

Oil fouling – Wet oily deposits leading to misfire and weak spark. Caused by oil leakage past piston rings or valve guides (4-stroke engine), or excess lubricant (2-stroke engine)

Overheating – A blistered white insulator and glazed electrodes. Caused by ignition system fault, incorrect fuel, or cooling system fault

Worn plug – Worn electrodes will cause poor starting in damp or cold weather and will also waste fuel

3   On H100 S models test the source coil by measuring the resistance between the terminal of the black/yellow wire and a good earth point on the crankcase; results should be as follows:

Black/yellow wire to earth   50 – 300 ohm

If the readings fall outside the limits given, the coil is faulty and the complete stator must be renewed. The comments made above apply equally to both models. An alternative is to find a competent auto-electrical company who may be able to rewind a faulty coil.

8   Condenser (capacitor): testing – H100 S

1   The condenser, more correctly referred to as the capacitor, is mounted in the generator stator plate. It is retained by a single screw and has the two wires of the low tension circuit soldered to the contact on its outer end. To remove it, lever up the wire retaining clamp, unsolder the connections and remove the single screw. The condenser can then be withdrawn. On reassembly do not forget to refit the cam lubricating wick and ensure that the wires are correctly soldered to the contact alone; if in doubt have the condenser refitted by an expert. It is easy to remove the stator plate so that the labour charge is reduced to a minimum.
2   If the engine is difficult to start, particularly when hot, or if misfiring occurs, it is likely that the condenser is at fault. The simplest way to check this is to remove the left-hand engine cover and to watch the contact breaker points as the engine is running. If a continuous stream of sparks can be seen as the points open and close, the condenser is faulty; do not, however, confuse this with the odd spark that occurs every now and then in normal use. If the contact faces are burnt and blackened, the condenser is almost certainly at fault. Unfortunately it can be tested only on a special electrical tester (note that it must be discharged to earth before any testing is attempted. With the bared end of a short length of wire held on the condenser body, touch the other end of the wire to the contact to earth it). In view of this, the simplest answer for most owners would be to renew the condenser if in doubt. Note, however, that replacement condensers are not listed separately for the H100 S; the only solution would seem to be the renewal of the stator assembly. It is recommended that the help is enlisted of the parts department staff of a good Honda Service Agent; if the original condenser is compared with those stocked for other models it should be possible to obtain a suitable replacement.

9   Diodes: location and testing – H100 S

1   The two diodes are small black plastic units rubber-mounted between the rectifier and the ignition HT coil on the frame top tube. Each is connected to the main loom via two spade terminals and can be unplugged easily on removal and refitting, after the petrol tank has been removed.
2   If the current flow between the generator, ignition switch and HT coil is interrupted, the diodes should be suspected. They can be tested using a multimeter set to its most sensitive resistance scale to check for continuity between the terminals, the polarity of which is indicated by the appropriate symbols being marked on the diode body. If in doubt, the black/white wire terminals are negative (-).
3   There should only be continuity (minimal resistance) from the negative (-) to the positive (+) terminal; much greater resistance should be measured when the meter probes are reversed to check for continuity in the other direction. If no resistance is found in either direction, or if heavy resistance is measured in both directions, the diode is faulty and must be renewed.

9.1 Location of diodes – H100 S

10   Ignition timing: checking – H100

1   Due to the nature of the ignition system used on this machine, it is not possible to adjust the ignition timing in any way. Routine checks of the ignition timing therefore, are not necessary, and should only be carried out when tracing the cause of a drop in performance. However, as the stator plate has slotted mounting holes there is a chance of its being fitted slightly out of line. It will be therefore necessary also to check the ignition timing whenever the stator plate is disturbed. It must be stressed however that the ignition timing can only be checked; if it is found to be incorrect the generator rotor and stator must be renewed.
2   The ignition timing can be checked only whilst the engine is running using a stroboscopic lamp and thus a suitable timing lamp will be required. The inexpensive neon lamps should be adequate in theory, but in practice may produce a pulse of such low intensity that the timing mark remains indistinct. If posssible, one of the more precise xenon tube lamps should be employed powered by an external source of the appropriate voltage.
3   Remove the left-hand outer cover. Connect the timing lamp to the machine as directed by the lamp's manufacturer. Start the engine and aim the lamp at the generator rotor. Increase engine speed to 3000 rpm and check that the fixed index mark on the crankcase wall is aligned exactly with the 'F' mark on the rotor.
4   In the event of the owner not having a tachometer with which to accurately measure the engine speed, the following method can be used, but must be regarded as only a rough guide. Start the engine and allow it to idle. Aim the timing lamp at the generator rotor. The fixed index mark should be approximately in the region of the 'T' mark. Slowly increase the engine speed. If the timing is correct, the 'F' mark will appear to move round to line itself up with the fixed index mark, and it will then remain reasonably steady at that point regardless of engine speed. It will be realised that this is not an accurate method of checking the ignition timing and should only be regarded as a rough indication until the machine can be taken to an authorised Honda dealer. A dealer should be able to check the ignition timing accurately, using a test unit fitted with a separate tachometer.
5   If there is any doubt about the ignition timing as a result of this check, take the machine to an authorised Honda dealer for an expert opinion. As already stated, there is no means of adjustment of the ignition timing on these machines, and renewal of the generator stator and rotor is the only solution to any fault. This is likely to prove rather expensive and an expert opinion would be a good idea to avoid wasting money.

10.2 A good quality timing lamp will be required to check ignition timing

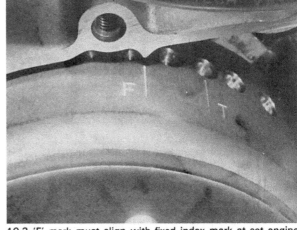

10.3 'F' mark must align with fixed index mark at set engine speed

11 High tension (spark plug) lead: examination

1   Erratic running faults and problems with the engine suddenly cutting out in wet weather can often be attributed to leakage from the high tension lead and sparking plug cap. If this fault is present, it will often be possible to see tiny sparks around the lead and cap at night. One cause of this problem is the accumulation of mud and road grime around the lead, and the first thing to check is that the lead and cap are clean. It is often possible to cure the problem by cleaning the components and sealing them with an aerosol ignition sealer, which will leave an insulating coating on both components.

2   Water dispersant sprays are also highly recommended where the system has become swamped with water. Both these products are easily obtainable at most garages and accessory shops. Occasionally, the suppressor cap or the lead itself may break down internally. If this is suspected, the components should be renewed.

3   Where the HT lead is permanently attached to the ignition coil, it is recommended that the renewal of the HT lead is entrusted to an auto-electrician who will have the expertise to solder on a new lead without damaging the coil windings.

12 Spark plug: checking and setting the gap

1   All models are fitted as standard with either an NGK BR7HS or ND W22FSR spark plug. In most operating conditions the standard plug should prove satisfactory. However, alternatives are listed to allow for varying altitudes, climatic conditions and the uses to which the machine is put. Consult a local authorised Honda dealer for advice before altering the plug specification from standard.

2   The correct electrode gap is 0.6 – 0.7 mm (0.024 – 0.028 in). The gap can be assessed using feeler gauges. If necessary, alter the gap by removing the outer electrode, preferably using a proper electrode tool. **Never** bend the centre electrode, otherwise the porcelain insulator will crack, and may cause damage to the engine if particles break away whilst the engine is running.

3   After some experience the sparking plug electrodes can be used as a reliable guide to engine operating conditions. See accompanying photographs.

4   It is advisable to carry a new spare sparking plug on the machine, having first set the electrodes to the correct gap. Whilst sparking plugs do not fail often, a new replacement is well worth having if a breakdown does occur.

5   Never overtighten a sparking plug otherwise there is risk of stripping the threads from the cylinder head, especially as it is cast in light alloy. A stripped thread can be repaired without having to scrap the cylinder head by using a 'Helicoil' thread insert. This is a low-cost service, operated by a number of dealers.

6   Before replacing a sparking plug into the cylinder head coat the threads sparingly with a graphited grease to aid future removal. Use the correct size spanner when tightening the plug otherwise the spanner may slip and damage the ceramic insulator. The plug should be tightened sufficiently to seat firmly on the sealing washer, and no more.

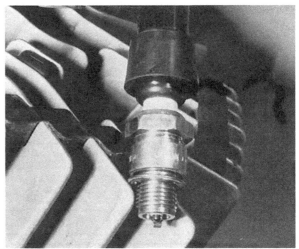

12.1 Spark plug must be checked regularly to maintain correct gap

Chapter 4 Frame and forks

For specifications and information relating to the H100 S II model, refer to Chapter 7

Contents

Specifications

Frame	Welded tubular steel	

Front forks

	H100	H100 S
Type	Oil damped telescopic	
Oil capacity – per leg	113.5 – 118.5 cc	80 – 86 cc
	(4.00 – 4.17 fl oz)	(2.82 – 3.03 fl oz)
Oil grade	Fork oil or ATF (Automatic Transmission Fluid)	
Fork spring free length	469.0 mm (18.46 in)	478.2 mm (18.83 in)
Wear limit	459.0 mm (18.07 in)	464.0 mm (18.27 in)
Fork stanchion bend – max	0.2 mm (0.008 in)	0.2 mm (0.008 in)

Rear suspension

Type	2 suspension units acting on pivoted fork	
Suspension unit type	Coil springs, hydraulically damped	Coil spring, gas damped
Spring free length	213.2 mm (8.39 in)	218.0 mm (8.58 in)
Wear limit	208.0 mm (8.19 in)	211.0 mm (8.31 in)

Torque settings

Component	kgf m	lbf ft
Fork crown nut	6.0 – 9.0	43 – 65
Handlebar clamp bolt	0.8 – 1.2	6 – 9
Fork top bolt	6.0 – 8.0	43 – 58
Bottom yoke pinch bolt	2.0 – 3.0	14 – 22
Front wheel spindle nut	5.5 – 6.5	40 – 47
Side stand pivot bolt	0.8 – 1.2	6 – 9
Footrest mounting bolt	2.5 – 3.5	18 – 25
Swinging arm pivot bolt	5.5 – 6.5	40 – 47
Shock absorber mounting nut	3.0 – 4.0	22 – 29
Rear brake torque arm	1.8 – 2.5	13 – 18
Rear wheel spindle nut	5.5 – 6.5	40 – 47

1 General description

The frame consists of a centre spine braced by triangulated steel tubes which support the pivoted rear fork mounting and seat rails. The engine hangs below the frame and does not form a part of it, which makes access for servicing work and engine removal very easy. Front suspension is by means of oil-damped telescopic forks. Rear suspension is by conventional coil sprung and oil or gas damped units which act on a pivoted fork.

2 Front fork removal: general

1 It is unlikely that the forks will require removal from the frame unless the fork seals are leaking or accident damage has been sustained. In the event that the latter has occurred, it should be noted that the frame may also have become bent, and whilst this may not be obvious when checked visually, could prove to be potentially dangerous.
2 If attention to the fork legs only is required, it is unnecess-

ary to detach the complete assembly, the legs being easily removed individually.

3 If attention to the steering head assembly is required it is possible to remove the lower yoke with the fork legs still in place, if desired. It should be noted, however, that this procedure is hampered by the unwieldy nature of the assembly, and it is recommended that the fork legs be removed prior to dismantling the steering head and fork yokes.

4 Before dismantling work can begin it will be necessary to arrange the machine so that the front wheel is raised clear of the ground. This is best done by lashing the rear of the machine down, either to a fixed object in the workshop or to a suitable weight.

5 The front wheel must now be removed as described in Section 3 of Chapter 5. If removal of the steering head assembly is required it is best to remove completely the speedometer cable and the front brake cable at this stage. If fork leg removal only is contemplated, it will suffice to tape these cables to the frame to keep them out of harm's way. Remove the mudguard by slackening and removing the four retaining bolts.

6 At this stage, if removal of the handlebars or steering head assembly is required, removal of the petrol tank is recommended, to prevent damage to the paintwork of the tank, and to allow easier access to the front fork components. See the relevant Section in Chapter 2 for details. At the very least, the tank should be covered with an old blanket or similar padding.

3 Front forks: removing the fork legs

1 Once the front wheel and mudguard have been removed as described in the previous Section, slacken the two fork top bolts, which project vertically through the top yoke, and the two bottom yoke pinch bolts. At this stage it is recommended that the handlebars are removed. While handlebar removal is not strictly necessary when removing the fork legs, it improves access a great deal, especially if the fork legs stick in the yokes and have to be drifted from position.

2 To remove the handlebars, the simplest method is to slacken and remove the four handlebar clamp bolts and to withdraw the two handlebar upper clamps. This allows the handlebar assembly to be lifted away from the top yoke and back clear of the steering head area. The handlebar assembly should then be rested on the frame top tube or on the padding covering the petrol tank. Once the handlebars are clear of the steering head area, remove the two front fork top bolts.

3 To remove the fork legs, slide each one down through the bottom yoke and put it to one side. If the legs prove stubborn, apply a small quantity of penetrating oil to the points at which they pass through the bottom yoke and to the top of the leg underneath the top yoke. Then free the chromed stanchion by attempting to rotate it. Do not clamp any tool to the stanchion during this operation as there is a considerable risk of scarring the chromed surface; these scars subsequently will allow severe rusting of the stanchion to take place. The stanchions should be rotated by hand only. If the stanchions still prove stubborn it will be necessary to use a hammer and a soft wooden drift to tap gently on the top of each leg in order to displace it. It is permissible gently to tap a screwdriver blade into the slot in the bottom yoke clamps, to open up the clamp very slightly and free the leg. The pinch bolts should be removed completely and great care must be taken to avoid overstressing the clamps.

4 Steering head assembly: removal and refitting

1 Prior to removal of the steering head assembly, the front wheel, front mudguard, and front fork legs must be removed as described in Sections 2 and 3 of this Chapter. First ensure tht the petrol tank has been either removed or suitably protected from accidental damage to the painted finish. Removal of the handlebars, if not already done, will be necessary also.

2 The official method of removing the handlebars is to remove all the handlebar controls, mirrors, and front flashing indicator lamps from the handlebars themselves before the handlebars are withdrawn. This method is only necessary if the handlebars are to be renewed. Provided sufficient care is taken, it will suffice to lift them clear of the top yoke and back on to the frame top tube or on to the padded petrol tank, clear of the steering head area, as described in Section 3.

3 For cases where the handlebars are to be renewed, the dismantling procedure is now given. Slacken the locknuts and unscrew the mirrors. Remove any cable ties securing cables or wiring to the handlebars. Slacken the two screws which fasten the two halves of the left-hand switch cluster and remove the screws. Carefully separate the two halves of the switch, withdraw them from the handlebar and allow the complete assembly to hang down by the side of the machine. Return the screws to their position in the switch casing for safekeeping. Slacken the two screws which fasten the two halves of the throttle twistgrip and the single pinch bolt which secures the front brake lever clamp, so that the twistgrip and front brake lever assemblies are free to slide on the handlebar. Slacken and remove the four handlebar clamp bolts and withdraw the two handlebar upper clamps. Remove the handlebars, moving them out to the left while sliding the twistgrip and front brake lever assemblies off the handlebar right-hand end.

4 At this stage, if the steering head bearings only are to be examined, it is possible to shorten the operation by proceeding directly to remove the top and the bottom yokes as described in paragraph 6 of this Section. This method will leave the complete headlamp casing assembly hanging in place without the need for disconnecting electrical wiring or for any further dismantling. It might, however, restrict access to the steering head bearings themselves or cause problems by getting in the way on reassembly. It is therefore recommended that the less experienced owner follows the longer, but ultimately easier, route now described. First the battery must be disconnected to prevent any accidental short circuits when the electrical components are disconnected. Slacken and remove the headlamp rim securing screws and withdraw the headlamp assembly. Disconnect the headlamp and parking lamp wires at their snap connectors and put the headlamp assembly to one side. Using a suitable pair of pliers, unscrew the knurled retaining ring at the upper end of the speedometer and tachometer cables and remove the cable. Disconnect all the electrical wiring at the snap connectors inside the headlamp casing and remove the main wiring leads from the switches, flashing indicator lamps, and from the main loom.

5 Unscrew the two side reflex reflectors, and slacken and then remove the two headlamp casing retaining bolts. Withdraw the headlamp casing/instrument panel assembly and put it to one side.

6 Remove the horn from the bottom yoke by unscrewing its single mounting bolt. Slacken and remove the single fork crown nut and withdraw the top yoke complete with the instrument panel (H100 S). This may need a careful tap with a soft-faced mallet to free it. Remove the two headlamp brackets.

7 Using a pin spanner or C-spanner, slacken and remove the steering head adjusting nut and then withdraw the top cone. Support the bottom yoke with the other hand while doing this. Some means of catching the steel balls must be devised, such as a large piece of rag wrapped around the bottom yoke, or a large plastic tray or container placed to catch the balls as they fall. If care is used, the balls in the top race will stay in place and enable you to watch the lower ones. There is a total of 42 steel balls of $\frac{3}{16}$ in diameter (No 6), 21 in each race. Withdraw the bottom yoke and place all the balls in a separate container to await examination and reassembly.

8 On reassembly, the bottom yoke must be installed first, as described in Section 5 of this Chapter. Temporarily slide the two fork legs into position in the bottom yoke, tightening the bottom yoke pinch bolts by just enough to retain the fork legs in

position. Slide the two headlamp brackets into place over the fork legs, ensuring that the upper and lower mounting rubbers on each are correctly located. Install the top yoke, then fit and hand-tighten the fork crown nut and the two fork leg top bolts. Check that the fork legs are correctly aligned and tighten the fork crown nut to a torque setting of 6.0 – 9.0 kgf m (43 – 65 lbf ft).

9   The reassembly procédure is then a straightforward reversal of the removal sequence described in this Section and in Sections 2 and 3 of this Chapter. Where such settings are given, use a torque wrench to tighten the relevant nuts and bolts. Note that the handlebar upper mounting clamps have a punch mark on one end which must be at the front when the clamp is fitted. There are two punch marks in the handlebar itself which must be aligned with the top faces of the lower clamps and the two handlebar front clamp bolts must be tightened to a torque setting of 0.8 – 1.2 kgf m (6 – 9 lbf ft) before the rear clamp bolts are tightened to the same torque setting. If no torque wrench is available, these bolts must be tightened just enough to retain the handlebars securely, or else the clamps will be overstressed, with a consequent risk of their cracking in use. When replacing the handlebar controls on the handlebars, note that there are punch marks in the handlebars with which the joint of each switch cluster or lever clamp must be aligned.

10   Pay careful attention to giving control cables and wiring leads easy smooth runs, this task being made much easier if the handlebar and controls have been correctly aligned using the punch marks just described. Refer to the colour-coding of the electrical wiring shown in the diagram at the back of this book when connecting the wiring. Do not forget to connect the battery again and to test all the circuits before using the machine on the road.

11   Before tightening the front wheel spindle nut, check that the front forks are correctly aligned, apply the front brake and push up and down on the handlebars to operate the front suspension and to ensure that all the fork components are settled in place. Using a torque wrench where applicable, tighten first the fork top bolts and then the bottom yoke pinch bolts, the mudguard mounting bolts and the front wheel spindle nut. Check that the steering head bearing adjustment is correct, as described in Section 5 of this Chapter.

12   Finally ensure that the punch marks on the headlamp casing are aligned with the reference marks on the headlamp brackets to give correct headlamp beam alignment and that all the controls are working smoothly and correctly before taking the machine out on the road.

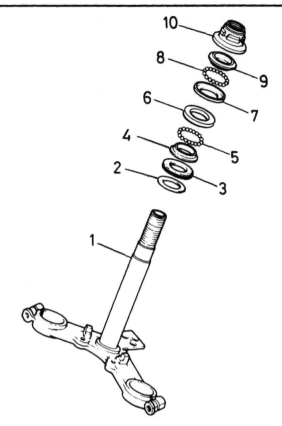

Fig. 4.1 Steering head assembly

1	Steering stem	6	Lower bearing cup
2	Washer	7	Upper bearing cup
3	Dust seal	8	Upper bearing balls – 21 off
4	Lower bearing cone	9	Upper bearing cone
5	Lower bearing balls – 21 off	10	Adjusting nut

5   Steering head bearings: examination, refitting and adjustment

1   Before commencing reassembly of the forks examine the steering head races. They are unlikely to wear out under normal circumstances until a high mileage has been covered. If, however, the steering head bearings have been maladjusted, wear will be accelerated. If, before dismantling, the forks had had a pronounced tendency to stick in one position when turned, often in the straight ahead position, the cups and cones are probably indented and need to be renewed.

2   Examine the cups and cones carefully; it is not necessary to remove the cups for this. The bearing tracks should be polished and free from indentations, cracks or pitting. If signs of wear are evident, the cups and cones must be renewed. In order for the straight line steering on any motorcycle to be consistently good, the steering head bearings must be absolutely perfect. Even the smallest amount of wear on the cups and cones may cause steering wobble at high speeds and judder during heavy front wheel braking. The cups are an interference fit on their respective seatings and can be tapped from position using a suitable long drift. The top cone, as already mentioned, is lifted away in the course of dismantling and is as easily replaced when rebuilding. The bottom cone, however, is a tight fit on the steering stem. Clamp the bottom yoke in a soft-jawed vice to hold it securely without damaging the painted finish and, using two screwdrivers or tyre levers, carefully lever the bottom cone away from its seating. Take great care not to damage the rubber dust seal and the washer which are situated underneath the

4.9 Punch marks in handlebars align with split clamps

bottom cone. Once the bottom cone is removed from its seating, it can be pulled easily off the steering stem. Carefully examine the rubber dust seal and renew it if worn, to prevent the entry of road dirt into the steering head bearings.

3   Replace the metal washer, rubber dust seal, and the bottom cone on the steering stem in that order. Position them carefully and use a hammer and a long tubular drift to tap the bottom cone on to its seating. Great care must be taken not to damage the bearing track of the bottom cone. Use a drift which will bear only on the inside shoulder of the cone to avoid such damage occurring. Tap the cups firmly on to their respective seatings in the frame using a suitably sized socket as a drift. Again take great care not to damage the bearing track surfaces.

4   The ball bearings themselves should be cleaned using paraffin. If found to be marked, chipped or discoloured in any way they should be renewed as a complete set.

5   On reassembly pack the bottom cone and top cup liberally with grease, and place the bearings in position, using the grease to hold them in place. Both races use twenty-one $\frac{3}{16}$ inch (No 6) steel balls. Note that this number will leave a gap for one more ball. This is intended, as some clearance is needed to prevent the balls skidding against one another and accelerating the rate of wear.

6   With the balls held in place as described, very carefully pack the bottom cup with grease and fit the bottom yoke. Support it with one hand and pack the area around the upper bearing with grease. Replace the top cone, ensuring that the balls do not become displaced in the process, and then thread the adjusting nut with its integral dust seal down into position. Using a pin spanner or C-spanner, tighten down the nut until it seats lightly. Do not overtighten it. Turn the nut back through $\frac{1}{8}$ turn from its lightly tightened position. This will serve as the basis for correct adjustment of the steering head bearings, which operation is now described in full. Remember that it can be carried out effectively only with the complete front fork assembly fitted, and should not be forgotten, therefore, during the rebuilding procedure.

7   The steering head bearings, when adjusted correctly, will have all traces of free play eliminated from them without being pre-loaded in any way. To check this, the machine must be placed on its centre stand with the front wheel raised clear of the ground by means of a box or other support underneath the engine. With the forks in the straight ahead position, grasp a fork lower leg in the area of the wheel spindle with one hand and attempt to push and pull the forks backwards and forwards. Any free play will be felt immediately by the fingers of the other hand between the bottom yoke and the frame around the lower steering head bearing. Any such free play must be just eliminated by slackening the fork crown nut above the top yoke and by tightening the adjusting nut immediately under the top yoke, using a C-spanner. To check for overtightened steering head bearings, position the machine as above, and push lightly on one handlebar end. The front forks should smoothly and easily fall away to the opposite lock. Any signs of stiffness or notchiness will be apparent immediately and should be removed by slackening off the adjusting nut. After adjusting the steering head bearings, tighten the fork crown nut to a torque setting of 6.0 – 9.0 kgf m (43 – 65 lbf ft) and recheck the bearing adjustment as described above to ensure that it has remained the same.

8   Note that if it is impossible to adjust the bearings correctly, it must be assumed that the steel balls or the bearing tracks are damaged in some way. The bearings should then be stripped for examination and renewal of the affected components.

6   Fork yokes: examination and renovation

1   To check the top yoke for accident damage, push the fork stanchions through the bottom yoke and fit the top yoke. If it lines up, it can be assumed the yokes are not bent. Both must also be checked for cracks. If they are damaged or cracked, fit new replacements.

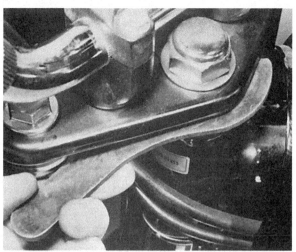

5.7a Steering head bearing adjustment should be made using a C-spanner

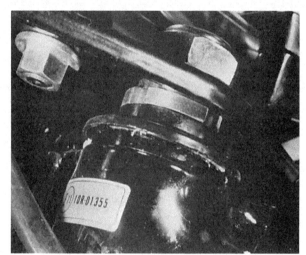

5.7b ... on the large slotted adjusting nut under the top yoke

7   Front forks: dismantling the fork legs

1   Due to the nature of the front forks employed on this machine, any maintenance or repair work to be carried out on the fork legs must be preceded by removal of the front wheel, front mudguard, and the fork legs themselves, as described in Sections 2 and 3 of this Chapter. Fortunately this is a simple operation which should rarely be required.

2   Using a suitably-sized Allen key, slacken and remove the threaded spring retaining plug, noting that it will be necessary to hold the stanchion while this is done. The easiest method is to clamp the leg in the bottom fork yoke, or failing this, to use a strap wrench. It is not advisable to clamp the stanchion in a vice because of the risk of damage due to scoring or overtightening. Use of this method is permissible, however, if soft jaw padding is used and if the vice is not overtightened. Be very careful when unscrewing the threaded plug as the fork spring is under tension and will push the plug clear with some force. It follows that firm hand pressure must be applied to the Allen key to counter this.

3   Slide out the fork spring, identifying the closer-spaced coils which must be uppermost when the spring is refitted. Invert the leg over a drain tray and leave it until the damping oil has

drained. Repeat the above procedure on the remaining leg, but note that each leg should be dismantled and reassembled separately to avoid interchanging components which would cause increased wear.

4   Wrap some rag around the lower leg and clamp the assembly in a vice, taking care not to overtighten and thus distort the lower leg. Using an Allen key, slacken and remove the damper bolt from the recessed hole in the bottom of the lower leg. This will often cause some difficulty because the bolt threads are coated in a locking compound and once slightly loose there is a tendency for the damper rod to turn in its seat. To overcome this problem a length of wooden dowel can be employed to hold the damper rod. Grind a coarse taper on one end of the dowel and insert it down the bore of the stanchion so that it engages in the recessed head of the damper rod.

5   It will now be necessary to push on the dowel to obtain grip on the damper. The dowel must not be allowed to turn, and to this end it is recommended that a self-locking wrench is clamped across its end to provide a handle. With an assistant restraining the damper rod, the bolt should now unscrew. If working alone, cut the dowel so that it lies about $\frac{1}{2}$ in lower than the top of the fully extended stanchion. The top bolt can be temporarily refitted to apply pressure to the dowel whilst the bolt is removed. On H100 S models, pull the stanchion out of the lower leg and invert it to tip out the damper rod and rebound spring. Remove the dust seal and circlip from the top of the fork lower leg and lever out the oil seal, taking care not to scratch or damage the lower leg.

6   On H100 models slide off the dust seal which is fitted around the top of the lower leg. Using a suitable pair of circlip pliers, remove the large circlip which retains the oil seal in the top of the lower leg. Remove the lower leg from the vice and clamp it at the wheel spindle lug with the leg and stanchion horizontal. Grasp the stanchion firmly, push it in to the lower leg as far as possible, and pull it sharply out. Repeat the process

7.6 Oil seal retaining circlip must be removed to allow separation of fork stanchion from lower leg

until the oil seal is driven out of its housing by the slide-hammer action of the stanchion and bushes. Once the stanchion is removed the oil seal and the top bush can be pulled off from the top. Ensure that the upper length of the stanchion is quite clean before doing this, to avoid unnecessarily damaging the seal bush. Invert the stanchion and tip out the damper rod and rebound spring.

7   Using a pair of pliers with a sharply-pointed nose, carefully remove the lower bush bottom circlip and slide the bush off the stanchion. The top circlip need not be removed unless work is to be done to the stanchion itself.

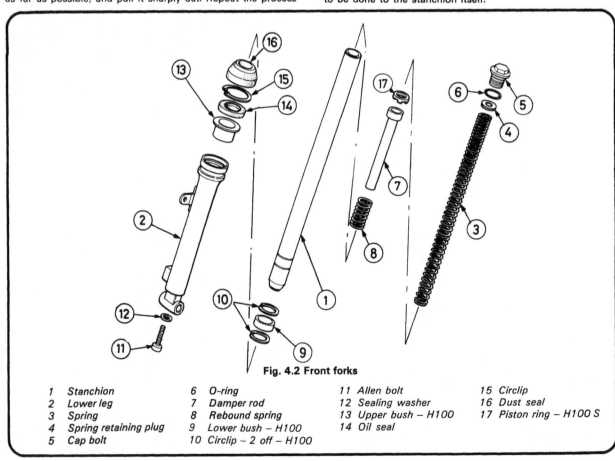

Fig. 4.2 Front forks

1	Stanchion	6	O-ring	11	Allen bolt	15	Circlip
2	Lower leg	7	Damper rod	12	Sealing washer	16	Dust seal
3	Spring	8	Rebound spring	13	Upper bush – H100	17	Piston ring – H100 S
4	Spring retaining plug	9	Lower bush – H100	14	Oil seal		
5	Cap bolt	10	Circlip – 2 off – H100				

8 Front forks: examination and renovation

1   Carefully clean all the fork components, removing all traces of rust, dirt, and old fork oil. Check that the oil control orifices in the damping rod are clear. Examine all the components, looking for excessive wear, hairline cracks or any other damage. The parts most likely to wear are the sealing lips of the dust seal and oil seal and the sliding surfaces of the stanchion, bushes, and the lower leg. Unfortunately no specifications are given by the manufacturer for any of these components and any assessment of wear will have to be by time-honoured trial and error methods. Any part found to be damaged, worn, or otherwise defective, must be renewed at once.

2   With all the components clean and dry, fit the top and bottom fork bushes temporarily back on the stanchion and secure the bottom bush with its lower circlip. Slide the stanchion assembly into the lower leg and clamp the wheel spindle lug firmly in a vice. Move the stanchion backwards and forwards in an attempt to feel for any play which may be present. As the lubricating film normally present between the sliding surfaces has been removed, any such free play should be immediately apparent. Repeat the test at several points, pulling the stanchion gradually out, until it is in the fully extended position. If any free play is felt, and the visual check revealed no serious signs of wear, take the fork leg assembly to an authorised Honda dealer for the various components to be compared with new parts. While it is inevitable that some free play will be present, especially in the fully extended position, a degree of experience will be necessary to assess whether any components will have to be renewed. Note that if this amount of wear has taken place, it will normally be evident due to areas of scoring on the component concerned. H100 S models are not fitted with separate bushes, but the above text can be applied to check for wear on the bearing surfaces of the stanchion and lower leg.

3   If any stiffness is encountered when moving the stanchion in the above test, it must be assumed that the stanchion is bent or distorted. To check this, remove the oil seal, the bushes, and both circlips from the stanchion and roll the stanchion on a dead flat surface. Any misalignment will be immediately obvious. If the forks have been removed to rectify damage incurred in an accident, this check will not, of course, be necessary. Bent or distorted stanchions may be straightened once if the tubing has not been cracked or creased by the impact. If the stanchion has been straightened previously, or if the tubing is damaged, the stanchion must be renewed. As a general rule, it is always best to err on the side of safety and fit new stanchions, especially since there is no easy means to detect whether the metal has been overstressed or fatigued. Remember that any attempt at straightening the stanchion must be made only by a qualified expert and should on no account be made by anyone without the equipment necessary and the skill and experience to use it.

4   If the stanchions have been found to be straight and unworn by the tests described, carefully examine the parts of them which are exposed when in position on the machine. Any signs of rust pitting must be removed. Deep pits can be filled with Araldite and then smoothed to match the original surface. It should be remembered that machines with fork stanchions pitted badly enough to damage the oil seals, and cause oil leakage, will be failed as unroadworthy when submitted for the annual MOT certificate. This is due to the reduction in damping efficiency which would result from such oil leakage. If pitting of the stanchions has taken place and is satisfactorily repaired, it is worth purchasing a pair of fork gaiters, available from any good motorcycle dealer, to cover the exposed stanchions and prevent this from happening in the future.

5   Check the lower legs carefully, looking for cracks around the mudguard mountings or the wheel spindle lugs. Any damage of this nature may be reclaimed by welding, but remember that this will inevitably destroy any chromed finish. Rechroming is likely to be so expensive that repairs become an uneconomic proposition, and the purchase of a new component is the cheaper course of action.

6   It is advisable to renew the oil seals whenever the forks are stripped, even if they appear to be in good condition. The seals are relatively cheap and their renewal at this stage will save the inconvenience of having to strip the forks at a later date, should oil leakage occur past the disturbed seals. They should be discarded and new ones fitted during assembly. The external dust seals should be checked for splits or wear but need only be renewed if damaged or worn. If fork gaiters are to be fitted, the dust seals should be left in place if possible to provide a second line of defence against dirt and corrosion.

7   The fork springs will take a permanent set after considerable usage and will need renewal if the fork action becomes spongy. The length of the fork springs should be checked against the figures given in the Specifications Section.

8   The damping action of the forks is governed by the viscosity of the oil in the fork legs, and the recommended types and capacities are given in the Specifications at the beginning of this Chapter. Note that when the fork is being topped up or the damping oil changed, slightly less oil will be needed. It is recommended that a piece of wire is used as a dipstick in these cases, and the oil level measured from the top of the fork stanchion. Given that the initial oil capacity is correct it will be possible to translate this to a known oil level, and subsequent refills and experiments in oil level changes can be made on this basis.

9   It is possible to increase or reduce the damping effect by using a different oil grade, and some owners may wish to experiment a little to find a grade which suits their particular application. It is advisable to consult a Honda Service Agent who will be able to suggest which oils may be used.

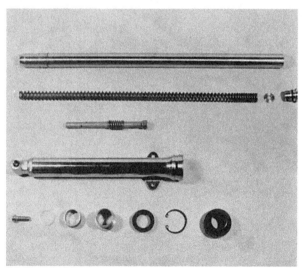

8.1 Front fork leg components

9 Front forks: reassembling the fork legs

1   All of the fork leg components should be completely clean and free from dust or oil prior to reassembly. Remember that the forks are in constant motion when in use, and any abrasive material will quickly wear away the surfaces between which it is trapped.

2   On H100 models fit the bottom bush upper retaining circlip into its groove in the stanchion, followed by the bottom bush which fits closely against it. Fit the lower retaining circlip into its groove and check that both circlips are properly fastened and located. They must not project beyond the bush or be able to move from their respective grooves.

3   Lightly lubricate the entire upper length of the fork stanchion with fork oil and slide the top bush down over the stanchion with its locating shoulder uppermost. Similarly fit the new oil seal, taking care not to damage the sealing lips. Place the rebound spring in position on the damper rod and use the fork spring or the length of dowel used during dismantling to push the damper rod down into the stanchion so that it projects from the stanchion lower end.

4   Lightly clamp the fork lower leg in a soft-jawed vice and feed the stanchion assembly into the lower leg, lightly lubricating the bushes with fork oil to assist in this. Check that the damper Allen bolt threads are clean and dry, coat them with Loctite and screw the Allen bolt into the damper rod. Lock the damper rod in the same way as used during dismantling and tighten the bolt to a torque setting of 0.8 – 1.2 kgf m (6 – 9 lbf ft).

5   Remove the fork spring or dowel rod and push the stanchion into the lower leg as far as possible. press the top bush down into the fork lower leg until its locating shoulder is firmly in position in the lower leg. Push the oil seal down into the lower leg by hand, using a small amount of grease on its outside diameter to assist this. Ensure that the seal is entered squarely into the bore of the lower leg. To drive the seal all the way into its housing a hammer and a long tubular drift must be used. The drift must be long enough to extend beyond the stanchion so that it can be tapped easily with the hammer, and it must bear only on the outer diameter of the seal. This will ensure that the seal is tapped squarely and evenly into its housing. Once the seal is in position, replace the large retaining circlip in its groove. Pack the area above the seal with grease to provide additional protection and slide the dust seal down over the stanchion and into place on the lower leg.

6   On H100 S models, tap the oil seal into place in the lower leg, using a hammer and a socket spanner which bears only on the seal outer edge. Ensuring that the seal enters squarely into its housing, tap it down until the circlip groove is exposed, then refit the circlip. Refit the rebound spring and piston ring to the damper rod, then insert the assembly into the stanchion and the stanchion into the lower leg as described above. Refit the damper rod Allen bolt and dust seal, also as described above.

7   Pull the stanchion out as far as possible and fit the fork spring, remembering that the closer-spaced coils should be at the top of the leg. Fit and tighten fully the spring retaining plug. Carefully fill the fork leg with the specified amount of fork oil. Remember not to accidentally invert the fork leg or allow any oil to spill. It is most important that both legs have exactly the same amount of oil in them if the suspension action is not to be impaired.

9.2a Fit bottom bush upper retaining circlip and slide top bush into place – H100

9.2b Slide bottom bush on to stanchion lower end – H100 ...

9.2c ... and secure bush with lower retaining circlip – H100

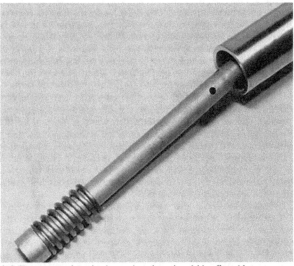

9.3 Damper rod and rebound spring should be fitted into stanchion

9.4a Lubricate bearing surfaces before fitting stanchion assembly into lower leg

9.4b Use thread locking cement on damper bolt threads

9.5a Grease exterior of seal to aid fitting

9.5b Secure seal with large circlip and pack grease around seal ...

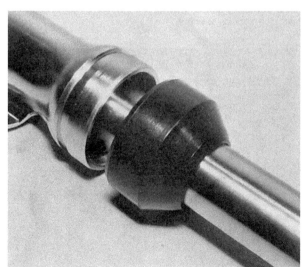

9.5c ... before fitting dust cover

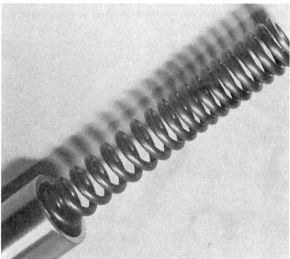

9.6a Close-spaced coils should be uppermost

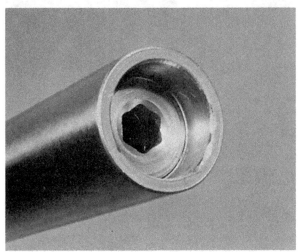

9.6b Threaded plug compresses fork spring – tighten carefully

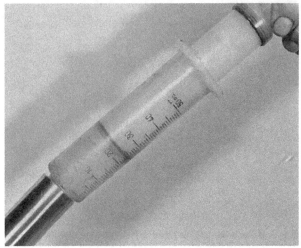

9.6c Use the same amount of fork oil in each fork leg

10 Front forks: refitting the fork legs

1   Replace the front forks by following in reverse the dismantling procedures described in Section 3 of this Chapter. Before fully tightening the front wheel spindle clamps and the fork yoke pinch bolts, bounce the forks several times to ensure they work freely and are clamped in their original settings. Complete the final tightening from the wheel spindle clamp upwards.

2   Do not forget to add the recommended quantity of fork damping oil to each leg before the bolts in the top of each fork leg are replaced if this has not already been done.

3   If the fork stanchions prove difficult to re-locate through the fork yokes, make sure their outer surfaces are clean and polished so that they will slide more easily. It is often advantageous to use a screwdriver blade to open up the clamps as the stanchions are pushed upward into position.

4   As the fork legs are fed through the yokes they will correct any slight misalignment between them should the steering head have been dismantled. Where appropriate, tighten the steering stem top nut and complete the fitting of the various ancillary components.

5   Before taking the machine out on the road, make a thorough test of the brakes and of the suspension action, and ensure that all nuts and bolts are securely fastened. Check the adjustment of handlebar controls and that all electrical equipment functions properly.

11 Frame assembly: examination and renovation

1   If the machine is stripped for a complete overhaul, this affords a good opportunity to inspect the frame for cracks or other damage which may have occurred in service. Check the top of the front downtube and the front of the top tube where it joins the steering head, the two points where fractures are most likely to occur. The straightness of the tubes concerned will show whether the machine has been involved in a previous accident.

2   Check carefully areas where corrosion has occurred on the frame. Corrosion can cause a reduction in the material thickness and should be removed by use of a wire brush and derusting agents.

3   If the frame is broken or bent, professional attention is required. Repairs of this nature should be entrusted to a competent repair specialist, who will have available all the necessary jigs and mandrels to preserve correct alignment. Repair work of this nature can prove expensive and it is always worthwhile checking whether a good replacement frame of identical type can be obtained at a reasonable cost.

4   Remember that a frame which is in any way damaged or out of alignment will cause, at the very least, handling problems. Complete failure of a main frame component could well lead to a serious accident.

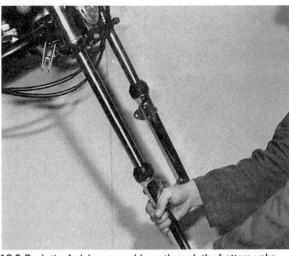

10.3 Push the fork leg assembly up through the bottom yoke

10.4a Tighten fork top bolt securely ...

10.4b ... the bottom yoke pinch bolt ...

10.4c ... and the mudguard mounting bolts

12 Rear suspension units: removal, examination and refitting

1   The models featured in this manual are equipped with coil spring suspension units using oil or gas-filled damper assemblies. The units are mounted at a steep angle to provide maximum rear wheel travel.

2   It is best to remove and attend to one unit at a time, because this will allow the machine to be supported by the remaining unit. Alternatively, place the machine on its centre stand so that the rear wheel is raised clear of the ground and the weight taken from the suspension units.

3   The units are retained by chromed dome-headed nuts at top and bottom mountings. Slacken and remove these and then remove the plain washer under each nut. Note that, on H100 models, it will be necessary also to remove the lifting handle on the left-hand side in order to allow removal of the suspension unit. This handle is retained by the left-hand suspension unit upper mounting nut at its front end and by a single bolt at its rear end. Withdraw the suspension units.

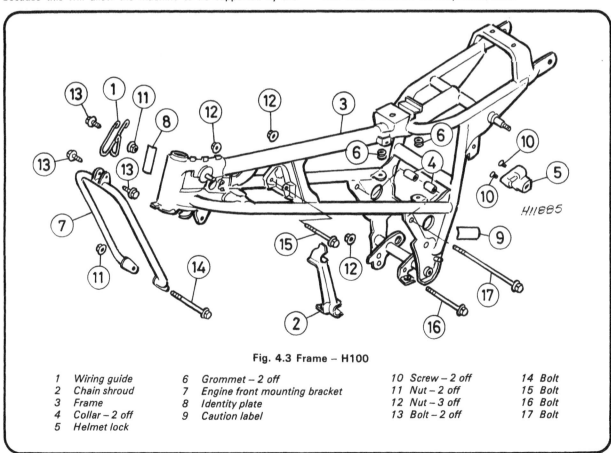

Fig. 4.3 Frame – H100

1	Wiring guide	6	Grommet – 2 off	10 Screw – 2 off	14 Bolt
2	Chain shroud	7	Engine front mounting bracket	11 Nut – 2 off	15 Bolt
3	Frame	8	Identity plate	12 Nut – 3 off	16 Bolt
4	Collar – 2 off	9	Caution label	13 Bolt – 2 off	17 Bolt
5	Helmet lock				

4   On H100 models set the spring adjuster ring to the softest position. Clamp the suspension unit's bottom mounting eye in a vice and enlist the aid of an assistant. With the assistant pulling the spring down, slip an open-ended spanner of suitable size over the slim locknut which secures the top mounting eye. Push a close-fitting metal bar through the top mounting eye and turn it to unscrew the top mounting from the damper shaft. Note that the assistant must keep the pressure firmly on the spring the whole time, and only release it gradually when the top mounting is removed. Withdraw the spring, noting the closer-spaced coils which must be uppermost on reassembly. With the spring removed, withdraw the locknut and rebound rubber from the damper shaft, and the adjuster ring from the damper body. On H100 S models, adopt a similar approach to remove the unit bottom mounting. If renewal of the rubber bushes in the top and bottom mountings is necessary, first push out the metal insert and then remove the rubber bush. Due to their lightweight construction they are very easy to remove and as easy to refit, although the use of liquid soap is recommended to ease the latter operation. All the suspension unit components should be carefully cleaned prior to examination.

5   Measure the free length of the spring, and renew it if it is below the service limit specified. The damper unit is sealed and cannot be dismantled. Its operation can be checked to some extent by compressing it and then releasing it. The unit should show significant damping effect on rebound (extension) but much less under compression. Any signs of leakage will necessitate renewal of the damper units as a pair.

6   If the suspension units appear to be in need of renewal, thought should be given to fitting an improved proprietary replacement pair. These may prove to be more expensive, but usually provide better control and durability. Most accessory shops will be able to advise and recommend the best make and type to use for any given application.

7   Reassembly of the units is a straightforward reversal of the dismantling sequence. When fitting the spring adjuster (H100), set it at its softest position to ease spring compression. When the units have been refitted set the spring adjusters at the required setting, ensuring that the position of each adjuster matches the other.

8   Remember that the closer-spaced spring coils must be uppermost and note that the mountings have a slot in them which must face inwards when the suspension units are refitted on the machine. Tighten the mounting nuts to a torque setting of 3.0 – 4.0 kgf m (22 – 29 lbf ft).

12.3a Rear suspension units are retained by dome-headed nuts which ...

12.3b ... must be removed to permit suspension unit to be withdrawn

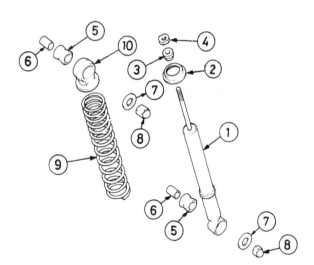

Fig. 4.4 Rear suspension unit – H100

1   Damper
2   Adjuster ring
3   Rebound rubber
4   Locknut
5   Bush – 2 off
6   Insert – 2 off
7   Washer – 2 off
8   Domed nut – 2 off
9   Spring
10 Top mounting eye

13 Swinging arm: removal, examination and refitting

1   Any wear in the swinging arm pivot bushes will cause imprecise handling of the machine, with a tendency for the rear wheel to twitch or hop. To check for any such wear, place the machine on its centre stand on level ground. Grasp the frame firmly with one hand and the fork end of the swinging arm with the other. Try to move the swinging arm from side-to-side in a

horizontal direction. Any play in the bushes should be immediately apparent at the swinging arm pivot. If there is any movement at all, the swinging arm must be removed and the bushes renewed.

2   To remove the swinging arm, the rear wheel must first be removed as described in Section 7 of Chapter 5. Slacken and remove the four bolts which retain the chaincase. Pull the chaincase lower half down at the front far enough to disengage it from its inner guide clip and withdraw it carefully from the machine. Push the chaincase upper half up at the front to disengage it from its inner guide clip and withdraw this carefully from the machine. Take great care not to damage the painted finish on either chaincase half. On H100 S models, remove the two retaining bolts and withdraw the chainguard.

3   On H100 models only, the chain need only be disconnected if its removal for inspection and cleaning or renewal is necessary. If such is the case, disconnect the chain at its connecting link and gently pull it away from the machine. If chain removal is not required, slacken the left-hand chain adjuster nut fully, slacken and remove the large sleeve nut which retains the rear sprocket assembly, and withdraw the sprocket assembly and chain adjuster from the swinging arm. Disengage the chain from the sprocket and loop the chain over the swinging arm fork end.

4   Slacken and remove the two nuts securing the suspension unit bottom mountings and pull the suspension units away from the mounting lugs on the swinging arm. Slacken, but do not remove, the two nuts securing the suspension unit top mountings and swing the two suspension units back and upwards. They can then be secured clear of the swinging arm areas with a length of string looped over the rear mudguard.

5   Slacken and remove the swinging arm pivot bolt securing nut. Withdraw the pivot bolt, using a hammer and a long metal drift to tap it out if necessary. Remove the swinging arm. Carefully clean the swinging arm, the pivot bolt, and the area of frame around the swinging arm pivot. Once all traces of oil, dirt and rust have been removed the parts concerned are ready for examination.

6   Carefully check all the components, looking for excessive wear, cracks, distortion, or serious corrosion. If the pivot bolt is at all worn it must be renewed, but damage to the frame or swinging arm may be repaired, depending on the nature of the damage. Any such repair work must be entrusted to an expert as described in Section 11 of this Chapter. If the swinging arm bushes have been found to be worn they must now be renewed. The type of bonded rubber bush used on this machine is notoriously difficult to remove and two possible means of achieving this are suggested.

7   The first method is shown in the accompanying photograph; it consists of a two-legged hydraulic puller adapted to suit this operation. Place a thick steel washer with an outside diameter the same as that of the bush outer sleeve against the bush. This will ensure that the pressure is applied in the correct place, against the sleeve, and not against the rubber which would only distort and eventually shear. Position the tool as shown in the photograph and press out the two bushes and central spacer, treating them as a single unit, far enough for the central spacer to be displaced. The tool can be then positioned on each swinging arm lug in turn, and the separate bushes pressed out individually. The central spacer is not fitted on H100 S models, so each bush can be removed and refitted individually.

8   The second method is exactly the same in principle, but uses much less expensive equipment which can be found in most workshops. Using the accompanying diagram as a guide, find a short length of thick-walled tube, the inside diameter of which is slightly larger than the outside diameter of the bushes, a large washer which will fit over the end of the tube, a smaller washer whose outside diameter is the same as that of the bush outer sleeves, and two high-tensile bolts with a nut for each. One of the bolts must be long enough to pass through the full width of the swinging arm and tube, while the other need only be long enough to pass through one lug of the swinging arm.

Assemble the tool as shown in the diagram with the long bolt passing through both bushes and the central spacer. Tighten down the nut, pressing out the two bushes and spacer as a single unit, far enough to permit the displacement of the spacer. Dismantle the tool, withdraw the spacer, and assemble the tool again, using the shorter bolt, on each of the swinging arm lugs in turn. Once the bushes have been freed, it becomes a comparatively simple task to press each one out.

9   If, due to corrosion between the mating faces of the bush and swinging arm lug, the bushes are reluctant to move, even using this method, it is recommended that the unit be returned to a Honda Service Agent whose expertise can be brought to bear on the problem. Note that, as a means of removal, attempting to drive the bushes out will probably prove unsuccessful, because the rubber will effectively damp out the driving force, and damage to the lugs may occur.

10 New bushes may be driven in using a tubular drift against the outer sleeve, or by reversing the removal operation, using the fabricated puller. Whichever method is adopted, the outer sleeve should be lubricated sparingly, and care must be taken to ensure that the bush remains square with the housing bore. The swinging arm must be properly supported whilst doing this. Ensure that the central spacer is located correctly before the bushes are in their final positions otherwise fitting the spacer will be impossible. Press the bushes in until their outer sleeves are flush with the outer edge of each lug, allowing the inner sleeve to project slightly.

11 On reassembly, liberally grease the pivot bolt and the inside diameter of the bushes and central spacer to minimise wear on the pivot bolt and to prevent corrosion. Replace the swinging arm in the frame, remembering to pass the left-hand fork end through the drive chain, and push the pivot bolt through from right to left. Fit and hand tighten the pivot bolt nut and washer. Untie the rear suspension units and lower them into place, fitting the bottom mounting eye of each over its respective lug on the swinging arm. Replace the suspension unit bottom mounting nuts and washers. Using a torque wrench, tighten all four suspension unit mounting nuts to a setting of 3.0 – 4.0 kgf m (22 – 29 lbf ft), and the swinging arm pivot bolt nut to a setting of 5.5 – 6.5 kgf m (40 – 47 lbf ft). It is important that the pivot bolt nut is only tightened once the swinging arm is in its normal working position. This minimises the distortion of the rubber bushes which would otherwise occur, thus preventing premature wear.

12 The remainder of the reassembly procedure is a straightforward reversal of the removal sequence. Before taking the machine out on the road, check the rear brake and chain adjustment and the rear suspension operation. Check also that all nuts and bolts are securely tightened and that all split pins, where fitted, are renewed and properly secured.

13.5a Slacken and remove pivot bolt nut, withdraw pivot bolt ...

13.5b ... and remove the swinging arm from the frame

13.6a Carefully examine bushes, mounting lugs and other areas of ...

13.6b ... the swinging arm itself for wear or damage

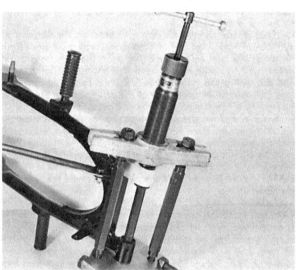

13.7 One suggested method of swinging arm bush extraction

13.11 Tighten swinging arm pivot bolt to recommended torque setting

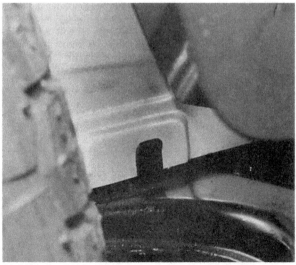

13.12 Ensure chaincase is correctly mounted on inner retaining clips – H100

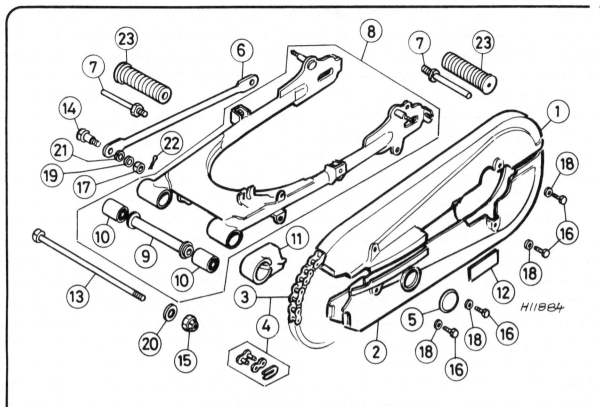

Fig. 4.5 Swinging arm and chaincase

1	Chaincase upper half – H100	6	Torque arm
2	Chaincase lower half – H100	7	Footrest bracket – 2 off – H100
3	Final drive chain	8	Swinging arm
4	Chain master link	9	Centre spacer – H100
5	Inspection plug – H100	10	Bush – 2 off
		11	Chain guide rubber

12	Caution label	19	Washer
13	Pivot bolt	20	Washer – 2 off – H100
14	Bolt	21	Spring washer
15	Nut	22	Split pin
16	Bolt – 4 off	23	Footrest rubber – 2 off
17	Nut		
18	Washer – 4 off		

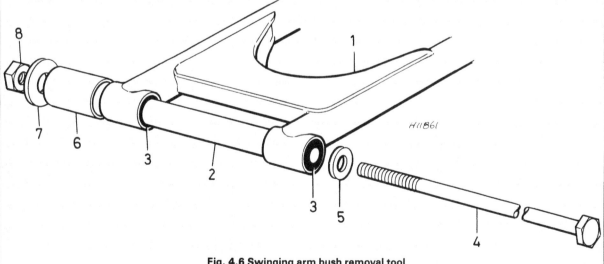

Fig. 4.6 Swinging arm bush removal tool

1	Swinging arm	5	Washer
2	Spacer – H100	6	Tube
3	Bush – 2 off	7	Washer
4	High tensile bolt	8	Nut

14 Footrests, stands and controls: examination and renovation

1   At regular intervals all footrests and stands and the brake pedal and gearchange lever should be checked and lubricated. Check that all mounting nuts and bolts are securely fastened, using the recommended torque wrench settings where these are given. Check that any securing split-pins are correctly fitted.
2   Check that the bearing surfaces at all pivot points are well greased and unworn, renewing any component that is excessively worn. If lubrication is required, dismantle the assembly to ensure that grease can be packed fully into the bearing surface. Return springs, where fitted, must be in good condition with no traces of fatigue and must be securely mounted.
3   If accident damage is to be repaired, check that the damaged component is not cracked or broken. Such damage may be repaired by welding, if the pieces are taken to an expert, but since this will destroy the finish, renewal is usually the most satisfactory course of action. If a component is merely bent it can be straightened after the affected area has been heated to a full cherry red, using a blowlamp or welding torch. Again the finish will be destroyed, but painted surfaces can be repainted easily, while chromed or plated surfaces can be only replated, if the cost is justified.

15 Speedometer and tachometer heads: removal, examination and reassembly

1   Before operations are described in detail, there are a few general observations to be made about these instruments. They must be carefully handled at all times and must never be dropped or held upside down. Dirt, oil, grease and water all have an equally adverse effect on them, and so a clean working area must be provided if they are to be removed.
2   The instrument heads are very delicate and should not be dismantled at home. In the event of a fault developing, the instrument should be entrusted to a specialist repairer or a new unit fitted. If a replacement unit is required it is well worth trying to obtain a good secondhand item from a motorcycle breaker in view of the high cost of a new instrument.
3   Remember that a speedometer in correct working order is a statutory requirement in the UK. Apart from this legal necessity, reference to the odometer readings is the most satisfactory means of keeping pace with the maintenance schedules.

H100
4   To remove the speedometer head, slacken and remove the single screw securing the headlamp rim to the headlamp casing. Withdraw the headlamp rim and unit assembly. While it can be left hanging by its connecting wires, it is much safer to disconnect these wires at their snap connectors and to put the headlamp assembly securely to one side to await refitting. Pull the black rubber warning lamp bulb holders down and away from the speedometer head, noting their positions to aid reassembly. Disconnect the four wires leading to the ignition switch. Slacken the knurled nut which secures the speedometer drive cable to the speedometer head and withdraw the cable. Carefully examine the large wire clip which retains the speedometer head in the headlamp casing, noting its exact position to ensure that it is replaced correctly on reassembly. Using a suitable pair of pliers, disengage the clip from the casing and from the speedometer head. Withdraw the speedometer head by pulling it upwards from above the headlamp casing. Using a small electrical screwdriver, free the plastic clips which retain the ignition switch and remove the switch.
5   If the instrument has been found to be faulty or is damaged it must be repaired only by an expert or renewed. No further stripping is possible or advisable for the private owner. See the general notes at the beginning of this Section.

6   Refitting is a straightforward reversal of the removal procedure. Carefully check for correct speedometer operation and that all electrical circuits have been properly connected. If necessary use the wiring diagram at the back of this manual as a guide.

H100 S
7   Remove the two bolts securing the headlamp. It is safest to disconnect the headlamp wires and remove the headlamp assembly, but is quicker to allow it to hang by the wires while work is completed on the instruments. Disconnect all wires leading to the instrument panel, slacken the two knurled nuts securing the speedometer and tachometer cables, and withdraw the cables. Slacken and remove the three screws which secure the bottom cover and remove the bottom cover. This exposes the two dome headed retaining nuts. Remove these and withdraw the complete instrument panel.
8   Using an electrical screwdriver, carefully free the plastic clips which retain the ignition switch and lift the switch upwards out of the panel. Taking great care not to invert the instrument panel any more than is necessary, free the four plastic clips, one at each corner, which secure the instrument mounting to the top cover and remove the top cover. Slacken and remove the two retaining screws to remove whichever instrument is required. Replacement is a straightforward reversal of the removal sequence.

16 Speedometer and tachometer drive cables: examination and maintenance

1   If the operation of the speedometer or tachometer becomes sluggish or jerky it can often be attributed to a damaged or kinked drive cable. Remove the cable from the machine by releasing the locating screw at the lower end and the knurled ring at the instrument head. Turn the inner cable to check for any tight spots. If the cable tends to snatch as it is rotated it is likely that the inner has become kinked and will require renewal.
2   Check that the outer cable is in sound condition with no obvious damage. The inner and outer cables are available separately where necessary. When fitting a new cable or refitting an old one, remove the inner and grease all but the upper six inches or so. This will ensure that the cable is adequately lubricated without incurring the risk of grease working up into the instrument head. Ensure that the cable is routed in a smooth path between the drive gearbox and instrument panel, avoiding any tight bends which would result in the cable becoming kinked.

15.4a Use a suitable pair of pliers to release speedometer cable

15.4b Note correct position of speedometer retaining clip before removing – H100

15.7 Remove two dome-headed nuts to release instrument panel

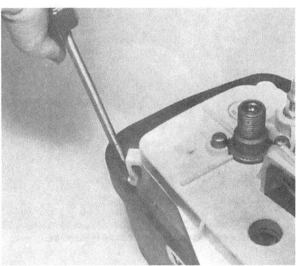

15.8a Use a suitable screwdriver to free the four plastic clips ...

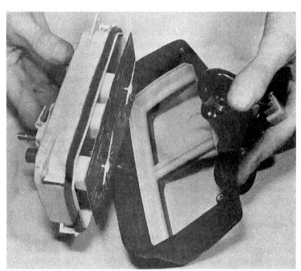

15.8b ... and remove the top cover

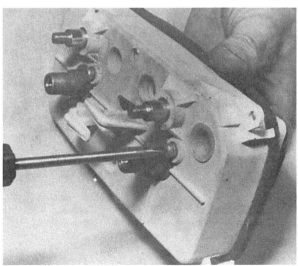

15.8c Remove the two retaining screws ...

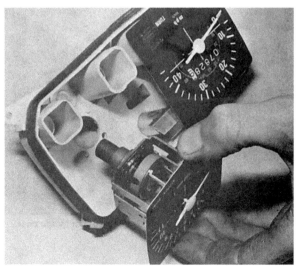

15.8d ... and withdraw the instrument

17 Speedometer and tachometer drives: location and examination

Speedometer – all models

1   The speedometer drive assembly is contained within the front wheel brake backplate and should be examined and repacked with grease whenever work is carried out on the wheel bearings or brake assembly.

2   To remove the main drive gear from its housing, place the brake backplate assembly, inner side uppermost, on a work surface. Remove the brake shoes and springs, as described in Section 5 of Chapter 5, and inspect the large dust seal for signs of damage and deterioration. To renew this seal, carefully lever it out of position using the flat of a screwdriver; great care must be taken not to damage the surrounding alloy casting. The new seal may be fitted after removal, examination and fitting of the drive gear assembly.

3   The main drive gear can be pulled out of the brake backplate housing; some difficulty may be experienced in withdrawal due to the gear being engaged with the worm drive gear and to the retentive qualities of the grease around the base of the gear.

4   Remove all old grease from the drive gear, brake backplate housing and worm gear by wiping the components with a clean rag. Inspect the gears for broken teeth and signs of excessive wear due to lack of lubrication. If it is considered necessary to renew the worm drive gear, then the brake backplate assembly should be returned to an official Honda Service Agent who may be able to remove and insert a new gear assembly.

5   Replace the main drive gear. Pack the housing with the recommended grease as this component is fitted. Fit the new dust seal, if required, into its recess in the brake backplate by pressing it into position evenly and squarely. The brake shoes and springs may now be refitted and the brake backplate assembly re-inserted into the wheel hub. Ensure that the speedometer drive tabs are aligned with the corresponding slots in the wheel hub boss.

Tachometer – H100 S

6   The tachometer drive consists of a worm mechanism driven by the crankshaft through the balancer idler gear and a driven gear mounted vertically in the right-hand outer cover. This mechanism is lubricated by the gearbox oil and is fully enclosed, therefore requiring no maintenance at all. Again any damage or malfunction can only be corrected by renewing the parts concerned. Inspect the mechanism whenever the right-hand outer cover is removed.

18 Steering lock: location, removal and refitting

1   The steering lock is situated at the front of the machine, on the bottom fork yoke. It is advisable to use it whenever the machine is left unattended, even for a short time.

2   To remove the lock, remove the countersunk screws holding the lock body to the bottom yoke bracket. Fitting the lock is the reverse procedure to removal. When the lock has been renewed, ensure a new key is obtained and carried with you when the machine is used.

17.2 Main speedometer drive gear can be lifted away

17.3 Use suitably shaped screwdriver to lever out large dust seal

17.5 Pack housing liberally with grease on assembly

Chapter 5 Wheels, brakes and tyres

For specifications and information relating to the H100 S II model, refer to Chapter 7

Contents

Specifications

Wheels

Type	Wire spoked, steel rims
Rim run out:	
Radial (max)	2.0 mm (0.08 in)
Axial (max)	2.0 mm (0.08 in)
Spindle warpage (max)	0.2 mm (0.008 in)

Brakes

Type	Single leading shoe drum
Drum ID	110.0 mm (4.33 in)
Wear limit	111.0 mm (4.37 in)
Lining thickness	4.00 mm (0.16 in)
Wear limit	2.00 mm (0.08 in)

Tyres

Front	2.50 – 18 4PR
Rear:	
H100	2.75 – 18 4PR
H100 S	2.50 – 18 4PR

Tyre pressures	Front	Rear
Solo	24 psi (1.75 kg/cm^2)	28 psi (2.0 kg/cm^2)
Pillion	24 psi (1.75 kg/cm^2)	32 psi (2.25 kg/cm^2)

Torque wrench settings

Component	kgf m	lbf ft
Wheel spindle nut (front and rear)	5.5 – 6.5	40 – 47
Sprocket mounting nut	2.7 – 3.3	19 – 23
Rear brake torque arm	1.8 – 2.5	13 – 18

1 General description

The wheels are of conventional wire-spoked construction, using chromed steel rims of 18 inch diameter. The front wheel is fitted with a ribbed tubed tyre and the rear with a block-tread tubed tyre. Brakes are mounted in full-width hubs at front and rear and are of the single leading shoe type. The front brake is actuated by a cable and the rear by a rod.

2   Front wheel: examination and renovation

1   Wire spoked wheels are often viewed as being prone to problems when compared to the increasingly popular cast alloy and composite types. Whilst this is true to some extent, it is also true that wire spoked wheels are relatively easy and inexpensive to adjust or repair. Spoked wheels can go out of true over periods of prolonged use and like any wheel, as the result of an impact. The condition of the hub, spokes and rim should therefore be checked at regular intervals.

2   For ease of use an improvised wheel stand is invaluable, but failing this the wheel can be checked whilst in place on the machine after it has been raised clear of the ground. Make the machine as stable as possible, if necessary using blocks beneath the crankcase as extra support. Spin the wheel and ensure that there is no brake drag. If necessary, slacken the brake adjuster until the wheel turns freely. In the case of rear wheels it is advisable though not essential, to remove the final drive chain.

3   Slowly rotate the wheel and examine the rim for signs of serious corrosion or impact damage. Slight deformities, as might be caused by running the wheel along a curb, can often be corrected by adjusting spoke tension. More serious damage may require a new rim to be fitted, and this is best left to an expert. Whilst this is not an impossible undertaking at home, there is an art to wheel building, and a professional wheel builder will have the facilities and parts required to carry out the work quickly and economically. Badly rusted steel rims should be renewed in the interests of safety as well as appearance. Where light alloy rims are fitted corrosion is less likely to be a serious problem, though neglect can lead to quite substantial pitting of the alloy.

4   Assuming the wheel to be undamaged it will be necessary to check it for runout. This is best done by arranging a temporary wire pointer so that it runs close to the rim. The wheel can now be turned and any distortion noted. Check for lateral distortion and for radial distortion, noting that the latter is less likely to be encountered if the wheel was set up correctly from new and has not been subject to impact damage.

5   The rim should be no more than 2.0 mm (0.1 in) out of true in either plane. If a significant amount of distortion is encountered check that the spokes are of approximately equal tension. Adjustment is effected by turning the square-headed spoke nipples with the appropriate spoke end. This tool is obtainable from most good motorcycle shops or tool retailers.

6   With the spokes evenly tensioned, any remaining distortion can be pulled out by tightening the spokes on one side of the hub and slackening the corresponding spokes from the opposite hub flange. This will allow the rim to be pulled across whilst maintaining spoke tension.

7   If more than slight adjustment is required it should be noted that the tyre and inner tube should be removed first to give access to the spoke ends. Those which protrude through the nipple after adjustment should be filed flat to avoid the risk of puncturing the tube. It is essential that the rim band is in good condition as an added precaution against chafing. In an emergency, use a strip of duct tape as an alternative: unprotected tubes will soon chafe on the nipples.

8   Should a spoke break a replacement item can be fitted and retensioned in the normal way. Wheel removal is usually necessary for this operation, although complete removal of the tyre can be avoided if care is taken. A broken spoke should be attended to promptly because the load normally taken by that spoke is transferred to adjacent spokes which may fail in turn.

9   Remember to check wheel condition regularly. Normal maintenance is confined to keeping the spokes correctly tensioned and will avoid the costly and complicated wheel rebuilds that will inevitably result from neglect. When cleaning the machine do not neglect the wheels. If the rims are kept clean and well polished many of the corrosion related maladies will be prevented.

3   Front wheel: removal and refitting

1   Place the machine securely on its centre stand on level ground and support the front wheel clear of the ground by means of a box, or wooden block placed under the engine. Using a suitable pair of pliers, slacken the knurled ring which retains the speedometer drive cable on the front brake backplate, and withdraw the cable. Ensure that it is secured out of harm's way. Slacken and remove the front brake adjusting nut from the end of the front brake cable. Gently pull the outer cable backwards, away from its retaining lug on the brake backplate, until the inner cable can be disengaged by pulling it through the slot in the lug. Pull the threaded end of the cable out of the nipple in the brake operating arm. If the wheel is to be removed for some time, the nipple must be replaced, with the adjusting nut, on the cable to ensure that they are not lost. Remove the split pin from the wheel spindle nut and unscrew the nut. Remove the spindle. It may be necessary to use a hammer and a soft metal drift gently to tap out the spindle. Withdraw the front wheel.

2   Refitting is a straightforward reversal of the removal procedure. Ensure that the speedometer drive tabs in the brake backplate are aligned with the corresponding slots in the wheel hub boss and align the groove in the brake backplate with the lug on the left-hand lower fork leg when replacing the wheel. Make sure that the wheel spindle is clean and completely free from corrosion, then grease it lightly before fitting to ease dismantling in the future. Push it through from right to left and hand tighten the spindle nut.

3   Connect the front brake cable again and apply the front brake lever hard to centralise the brake shoes and backplate with the brake drum. While keeping firm pressure on the brake lever, tighten the wheel spindle nut to a torque setting of 5.5 – 6.5 kgf m (40 – 47 lbf ft). Fit a new split pin and spread its ends securely. Replace the speedometer cable, spinning the wheel if necessary to assist the inner cable to engage with the drive mechanism. Securely tighten the knurled retaining ring. Adjust the front brake as described in Section 5 of this Chapter. Finally check for free front wheel rotation and that the speedometer and front brake work properly.

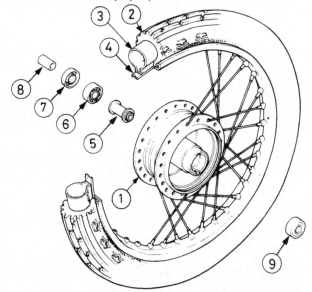

Fig. 5.1 Front wheel

1 Hub	6 Right-hand wheel bearing
2 Tyre	7 Oil seal
3 Inner tube	8 Right-hand spacer
4 Rim tape	9 Left-hand wheel bearing
5 Centre spacer	

3.1 Use a pair of pliers to release speedometer cable retaining ring

3.2a Ensure speedometer drive tabs align with slots in hub (arrowed)

3.2b Align lug on lower fork leg with groove in brake backplate

3.2c Do not omit spacer on right-hand side of hub

3.3a Tighten spindle nut to recommended torque setting and fit new split pin

3.3b Adjust front brake and replace speedometer cable

4 Front wheel bearings: removal, examination and refitting

1 In order to gain access to the wheel bearings, the front wheel must first be removed as described in Section 3 of this Chapter. Withdraw the brake backplate and put it to one side. Pull the spacer out of the dust seal on the right-hand side of the hub. To remove the dust seal alone, heat the end of an old flat-bladed screwdriver and bend the tip into a slightly curved shape with no sharp edges. This will give a useful tool for levering out the dust seal without damaging its sealing lip or the seal housing. If required, the dust seal and spacer may be driven out with the right-hand bearing.

2 To remove the bearings, support the wheel firmly on two wooden blocks placed as close to the hub centre as possible to prevent distortion and ensuring that the blocks are tall enough to provide space for the bearings. Place the end of a small flat-ended drift against the upper face of the lower bearing and tap the bearing downwards out of the wheel hub. The spacer located between the two bearings may be moved sideways slightly in order to allow the drift to be positioned against the face of the bearing. Move the drift around the face of the bearing whilst drifting it out of position, so that the bearing leaves the hub squarely. The end spacer and dust seal will be driven out along with the bearing.

3 With the one bearing removed, the wheel may be lifted and the spacer withdrawn from the hub. Invert the wheel and remove the second bearing, using a similar procedure to that used for the first.

4 Wash the bearings thoroughly in clean petrol to remove all traces of the old grease. Check the bearing tracks and balls for wear or pitting or damage to the hardened surfaces. A small amount of side movement in the bearing is normal but no radial movement should be detectable. Check the bearings for play and roughness when they are spun by hand. All used bearings will emit a small amount of noise when spun but they should not chatter or sound rough. If there is any doubt about the condition of the bearings they should be renewed.

5 Carefully clean the bearing recess in the hub and the centre space of the hub. All traces of the old grease, which may be contaminated with dirt, must be removed. Examine the dust seal and renew it if any damage or wear is found.

6 Before refitting the bearings pack them with high melting point grease. This applies equally to the original bearings, if re-used, and to new ones, if the originals are to be replaced. With

the wheel firmly supported on the two wooden blocks, tap a bearing into place in the hub, noting that the sealed surface must face outwards. Use a hammer and a tubular metal drift or socket spanner, which bears only on the outer race of the bearing, to drive the bearing into position. If the inner race or sealed surface of the bearing are used to drive it into place, severe damage will be done to the bearing due to the high side loadings thus imposed.

7 Once one bearing has been installed, invert the wheel, fit the central spacer, and pack the remaining space no more than $\frac{2}{3}$ full of grease. This is important as although some grease must be present, it will expand when hot and if too much grease is in the centre space, it will force its way past the seals and out on to the brake components. Once the grease is packed in, fit the second bearing in the same manner as the first. The dust seal in the right-hand side of the hub can be pressed into position by hand. The wheel spacer should then be greased and pushed into the dust seal.

8 Once the bearings are fitted remove any surplus grease from the brake drum and the hub exterior surface. Replace the front wheel in the forks as described in Section 3 of this Chapter.

4.7a Do not omit central spacer

4.7b Bearing sealed surface must face outwards

4.7c Tap bearings into place using socket as a drift

5   Front brake: examination, renovation and adjustment

1   The front brake assembly complete with the brake backplate can be withdrawn from the front wheel hub after removing the front wheel from the forks. With the wheel laid on a work surface, brake backplate uppermost, the brake backplate may be lifted away from the hub. It will come away quite easily, with the brake shoe assembly attached to its back.

2   Examine the condition of the brake linings. If they are thin or unevenly worn, the brake shoes should be renewed. The linings are bonded on and cannot be supplied separately. The linings are 4 mm (0.2 in) thick when new and should receive attention when worn to the wear limit thickness of 2 mm (0.1 in).

3   If fork oil or grease from the wheel bearings has badly contaminated the linings, the brake shoes should be renewed. There is no satisfactory way of degreasing the lining material.

4   Examine the drum surface for signs of scoring or oil contamination. Both of these conditions will impair braking efficiency. Remove all traces of dust, preferably using a brass wire brush, taking care not to inhale any of it, as it is of an asbestos nature, and consequently harmful. Remove oil or grease deposits, using a petrol soaked rag.

5   If deep scoring is evident, due to the linings having worn through to the shoe at some time, the drum must be skimmed on a lathe, or renewed. Whilst there are firms who will undertake to skim a drum whilst fitted to the wheel, it should be borne in mind that excessive skimming will change the radius of the drum in relation to the brake shoes, therefore reducing the friction area until extensive bedding in has taken place. Also full adjustment of the shoes may not be possible. If in doubt about this point, the advice of one of the specialist engineering firms who undertake this work should be sought.

6   Note that it is a false economy to try to cut corners with brake components; the whole safety of both machine and rider being dependent on their good condition.

7   Removal of the brake shoes is accomplished by folding the shoes together so that they form a 'V'. With the spring tension relaxed, both shoes and springs may be removed from the brake backplate as an assembly.

8   Before fitting the brake shoes, check that the brake operating cam is working smoothly and is not binding in its pivot. The cam can be removed by withdrawing the retaining bolt on the operating arm and pulling the arm off the shaft. Before removing the arm, it is advisable to mark its position in relation to the shaft, so that it can be relocated correctly. The shaft and arm should be already marked with a manufacturer's punch mark to indicate the correct relative positions of the two

components. Lightly grease both the shaft and the faces of the operating cam and pivot prior to reassembly.

9   Before refitting existing shoes, roughen the lining surface sufficiently to break the glaze which will have formed in use. Glasspaper or emery cloth is ideal for this purpose but take care not to inhale any of the asbestos dust that may come from the lining surface.

10   Fitting the brake shoes and springs to the brake backplate is a reversal of the removal procedure. Some patience will be needed to align the assembly with the pivot and operating cam whilst still retaining the springs in position; once they are correctly aligned though, they can be pushed back into position by pressing downwards in order to snap them into position. Do not use excessive force, or there is risk of distorting the brake shoes permanently.

11   Adjusting the front brake is best accomplished with the front wheel free to rotate. Spin the wheel and carefully screw the adjusting nut down until you hear a rubbing sound which shows that the brake shoes are lightly in contact with the drum surface. Turn the adjuster nut back by $\frac{1}{2}$ – 1 turn until the noise stops. Spin the wheel and apply the brake hard once or twice to settle the brake. Check that the wheel is still free to rotate and that the adjustment has remained the same. This setting should give you 20 – 30 mm ($\frac{3}{4}$ – $1\frac{1}{4}$ inch) free play measured at the extreme tip of the handlebar brake lever.

5.4 Examine surface of brake drum for scoring or contamination

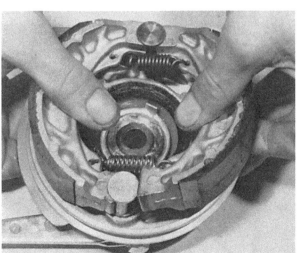

5.7 Method of removing brake shoes for inspection or renewal

5.8 Align punch marks to ensure operating arm is replaced correctly on brake cam

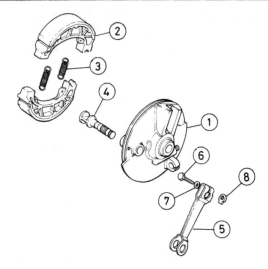

Fig. 5.2 Front brake assembly

1 Backplate	5 Operating lever
2 Shoe - 2 off	6 Pinch bolt
3 Return spring - 2 off	7 Washer
4 Operating cam	8 Nut

6 Rear wheel: examination and renovation

1 Place the machine on its centre stand and ensure that the rear wheel is quite free to revolve.
2 Since the front wheel and the rear wheel are identical in construction, the technique for examination and renovation is exactly the same as that given in Section 2. Refer to that Section when inspecting the rear wheel, but pay extra attention to spoke tension as the spokes are under more stress in the rear wheel.

7 Rear wheel: removal and refitting

H100

1 Place the machine on its centre stand on level ground to raise the rear wheel clear of the ground. Unscrew the rear brake adjusting nut from the end of the brake rod and apply the rear brake pedal to bring the brake rod forward and clear of the nipple in the operating arm. If the wheel is to be removed for some time, the retaining spring, nipple, and the adjusting nut should be replaced on the brake rod for safekeeping. Remove the split pin securing the torque arm retaining nut and slacken and remove the nut, the two washers, and the torque arm. Replace the nut, washers, and splitpin on the bolt for safe-keeping.
2 Remove the split pin from the wheel spindle nut and slacken, then remove, the spindle nut itself and the chain adjuster. Withdraw the spindle, using a hammer and a soft metal drift if necessary gently to tap it out. Remove the spacer situated between the brake backplate and the swinging arm and pull the wheel to the right, to disengage it from the vanes of the cush-drive assembly, and then backwards away from the machine. Tilt the machine carefully to the left to provide clearance for the wheel if necessary.
3 Refitting is a direct reversal of the removal procedure. Replace the brake backplate in the hub and fit the wheel back into the machine, ensuring that the vanes of the sprocket carrier engage fully with the cush-drive rubbers. Fit the spacer between the brake backplate and swinging-arm right-hand fork end. Ensure that the wheel spindle is completely clean and free from

corrosion, then grease it lightly to ease dismantling in the future. Push it through from left to right and replace the right-hand chain adjuster and the spindle nut. Hand tighten the spindle nut, check that the right-hand chain adjuster is aligned with the same swinging-arm index mark as on the left-hand side, and connect up the torque arm and brake rod again.
4 Apply the rear brake pedal hard to centralise the brake shoes and backplate on the drum and tighten the spindle nut to a torque setting of 5.5 – 6.5 kgf m (40 – 47 lbf ft). Fit new split pins to the wheel spindle and to the torque arm retaining nut and spread their ends securely. Complete rear brake and stop lamp switch adjustment as described in Section 9 of this Chapter. Finally check that the rear wheel is free to rotate, that the back brake works correctly and that all nuts and bolts are securely fastened.
5 Note that while chain adjustment is not altered by the action of removing the rear wheel, it is essential that accurate wheel alignment is preserved by ensuring that the index marks on the chain adjusters are both lined up with the same reference mark on the swinging arm. This should be carefully checked before tightening the wheel spindle nut.

H100 S

6 Place the machine on its centre stand on level ground to raise the rear wheel clear of the ground, then disconnect the rear brake rod by unscrewing the adjuster nut and removing the rod from the operating arm. Remove the split pin securing the torque arm retaining nut and slacken and remove the nut, two washers and the torque arm. Slacken and remove the rear spindle nut and withdraw the spindle. It may be necessary to use a hammer and soft metal drift gently to tap out the spindle. Slide the rear wheel forwards and disengage the chain from the rear sprocket. Withdraw the rear wheel, tilting the machine to one side to provide clearance if necessary. Note that it is not necessary to disconnect the chain unless required. If this is the case, find the connecting link, remove its spring clip using a pair of pliers and withdraw the connecting link. Try not to let the ends of the chain pick up any dirt or debris from the floor.
7 On refitting, do not forget to engage the chain back on the rear sprocket before fitting the wheel spindle. Ensure that the reference marks on the chain adjuster face the correct way up, so that they can be aligned with the swinging arm index marks when chain adjustment is required. Hand tighten the spindle nut, connect the brake torque arm and brake rod again, and complete chain adjustment and rear brake adjustment. Apply the rear brake pedal hard to centralise the shoes on the drum and tighten the spindle nut to a torque setting of 5.5 – 6.5 kgf m (40 – 47 lbf ft). Finally check that the rear wheel is free to rotate, that the brake and stop lamp switch work properly and that all nuts and bolts are securely fastened.

7.1 Release brake rod and torque arm prior to wheel removal

7.2a Push out wheel spindle and withdraw spacer

7.2b Sprocket (and chain) stay in place during wheel removal – H100

7.4 Always fit new split pins to brake torque arm and wheel spindle nuts

8 Rear wheel bearings: removal, examination and refitting

1   In order to gain access to the rear wheel bearings, the rear wheel must be first removed as described in Section 7 of this Chapter. Withdraw the brake backplate and put it to one side.
2   While the front and rear hubs are dissimilar in appearance the procedures for removal, examination, and refitting of the rear wheel bearings are exactly the same as those given in Section 4 for the front wheel. Refer to that Section, therefore, when working on the rear wheel.
3   While the rear wheel is removed it is advisable to check on the condition of the O-ring and cush drive rubbers set in the left-hand side of the hub. Any sign of damage to these items means that they must be renewed. If the transmission is excessively snatchy at low speeds, the cush drive rubbers should be inspected and renewed if necessary. Compare them with new components to assess their condition. If renewal is necessary, refer to Section 10.
4   Directions for refitting the rear wheel are given in Section 7 of this Chapter.

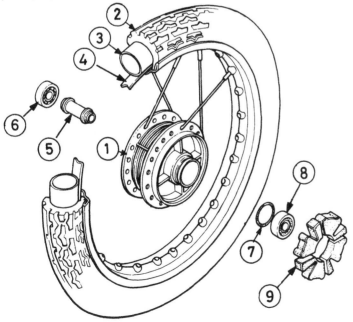

Fig. 5.3 Rear wheel – H100

1   Hub
2   Tyre
3   Inner tube
4   Rim tape
5   Centre spacer
6   Right-hand wheel bering
7   O-ring
8   Left-hand wheel bearing
9   Cush drive rubbers

8.3a Cush drive rubbers must be examined ...

8.3b ... and renewed if necessary – H100

9 Rear brake: examination, renovation and adjustment

1   The rear brake on this machine is identical in construction to that fitted to the front wheel. Once the rear wheel has been removed as described in Section 7 of this Chapter, the rear brake can be examined and attended to following the instructions given in Section 5.

2   Rear brake adjustment is made at the adjuster nut at the extreme end of the brake operating rod. Turn the nut clockwise to tighten the brake up, reducing free play at the brake pedal tip. Honda specify that the brake pedal should have 20 – 30 mm ($\frac{3}{4}$ – $1\frac{1}{4}$ inch) free play measured at its tip. Always spin the rear wheel to check that it rotates freely and that the brake is not binding once adjustment has been made. Remember also that if the brake adjustment has been altered significantly the stop lamp switch will have to be adjusted to suit. Turn the plastic retaining sleeve nut so that the stoplight comes on first as the brake pedal has taken up its free play and is starting to engage the brake.

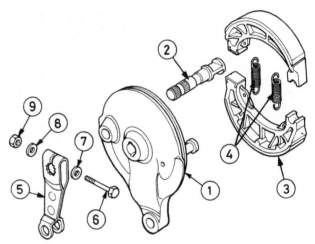

Fig. 5.4 Rear brake assembly

1	Backplate	6	Pinch bolt
2	Operating cam	7	Washer
3	Shoe - 2 off	8	Washer
4	Return spring - 2 off	9	Nut
5	Operating arm		

9.1a Rear brake assembly can be lifted away from hub ...

9.1b ... and is identical to front brake

10 Rear sprocket and cush drive assembly: removal, examination and refitting

H100

1   This machine is fitted with a transmission shock absorber in the form of a cush drive assembly built into the rear wheel. This serves to damp out transmission shock loads, thus producing a smoother ride and, in conjunction with the chaincase full enclosure, greatly extended chain life.

2   The cush drive employed consists of a cast alloy sprocket carrier, to which the rear sprocket is secured by four studs and nuts, riding on a ball journal bearing. This bearing is mounted on a sleeve which is secured to the swinging arm, independently of the rear wheel spindle, by a large sleeve nut. This permits the sprocket and carrier assembly to remain in place, undisturbed, during rear wheel removal. Large vanes cast in the sprocket carrier engage in slots moulded in four large synthetic rubber blocks which are held in four separate compartments in the left-hand side of the wheel hub. It is the rubber blocks which provide the damping effect and, therefore, are under some stress when in use. There is no maintenance to be carried out on them, but they should be inspected as described in Section 8 of this Chapter whenever the rear wheel is removed. If damaged or excessively worn, they must be renewed.

3   In order to gain access to the sprocket and carrier, the rear wheel must be first removed as described in Section 7 of this Chapter. Once the wheel is withdrawn, slacken and remove the four bolts which fasten the chaincase. Pull the chaincase lower half down at the front far enough to disengage it from its inner guide clip and carefully remove the chaincase half. Push the chaincase upper half up at the front, again to free it from its inner guide clip, and withdraw this from the machine.

4   If the chain is to be removed for inspection it should be disconnected at the connecting link and carefully pulled off the sprockets. If this is not the case, it can be left in place. Slacken and remove the large sleeve nut retaining the sprocket carrier and withdraw the left-hand chain adjuster and the sprocket carrier. If the chain is to be left in place, disengage it from the rear sprocket and loop it over the fork end of the swinging arm. Pull the spacer out of the left-hand side of the sprocket carrier and tap the carrier mounting sleeve out to the right. Slacken and remove the four sprocket retaining nuts and withdraw the rear sprocket.

5   The bearing and oil seal in the sprocket carrier should be attended to following the instructions given in Section 4 of this Chapter. Note that the sealed surface of the bearing should face

outwards, to the left, on reassembly. Wear on this bearing will rapidly accelerate chain wear, due to the rear sprocket being allowed to rotate unevenly.

6   Check the condition of the sprocket teeth. If they are hooked, chipped or badly worn, the sprocket must be renewed. It is, however, bad practice to renew one sprocket on its own, and so if the rear sprocket is considered badly enough worn to warrant renewal, the drive sprocket and chain should also be renewed. If only one new component is fitted, rapid wear will result from the running together of worn and unworn parts. This will necessitate the renewal of both sprockets and the chain after a very short time.

7   Reassemble the sprocket carrier and use a torque wrench to tighten the sprocket retaining nuts to a torque setting of 2.7 – 3.3 kgf m (19 – 23 lbf ft). It is recommended that a thread locking cement is used on the threads of these nuts, to prevent their coming undone. Complete the reassembly by reversing the instructions given in this Section and in Section 7 of this Chapter.

H100 S

8   This model is fitted with a transmission shock absorber of a simpler type, in which pins at the rear of the sprocket pass into bonded rubber bushes pressed into the wheel hub. To remove the sprocket, remove first the rear wheel as described in Section 7 of this Chapter. Remove the large circlip which secures the sprocket and pull away the sprocket. If the sprocket mounting pins are corroded into the cush drive bushes, remove the four nuts and lift away the sprocket, leaving the pins in place. Check the sprocket as described above.

9   The bonded rubber bushes are an extremely tight fit in the hub. If they have become compacted or the bonded bush centres are damaged (as shown by the locating pins jamming in position), they should be renewed. It is extremely unlikely that this operation can be performed at home due to the tight fit on the bushes. In all probability, any attempt to dislodge them will result in the inner metal sleeve tearing out of the rubber, making subsequent removal difficult. For this reason it is suggested that the wheel should be taken to a Honda Service Agent who will have the equipment necessary to extract the old bushes and fit the new ones.

10   Reassembly is a straightforward reversal of the dismantling procedure. It is important that the recesses in the rear of the sprocket are engaged correctly by the milled flats on each cush drive pin. Tighten the sprocket retaining nuts to a torque setting of 2.7 – 3.3 kgf m (19 – 23 lbf ft). Refit the rear wheel.

10.2 Cast alloy vanes engage in slots in rubber blocks

10.4 Sprocket can be withdrawn when rear wheel, chainguard and chain have been removed

10.7a Use socket as a drift to tap bearing home

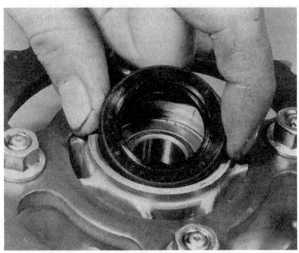

10.7b Dust seal must be renewed if necessary

10.7c Carrier spindle is fitted from right to left

10.7d Spacer is fitted into dust seal from left-hand side

11 Final drive chain: examination, adjustment and lubrication

1   The final drive chain employed on the H100 enjoys a relatively easy life due to the fitting of a fully-enclosed chaincase which will prevent water and road dirt from getting into the chain as would be the case with the exposed chain on the H100 S. This does not, however, mean that an 'out of sight, out of mind' attitude can be adopted. The chain will still require constant and regular lubrication and adjustment if it is to last as long as possible.

2   While routine lubrication and adjustment can be made with the chain in place, more extensive lubrication and inspection will require the removal of the chain. To accomplish this, first acquire a scrap length of 428 ($\frac{1}{2}$ x $\frac{5}{16}$ inch) size chain, 108 links long, perhaps from the dustbin of a local motorcycle dealer. Slacken and remove the two bolts retaining the chaincase lower half and withdraw the chaincase half (H100). Be careful not to damage the paintwork of the chaincase. Spin the back wheel until the chain connecting link comes into view. Disconnect the chain at this link and temporarily connect up the scrap length. By pulling on the free end of the original

chain, the scrap length will pass into the chaincase and around both sprockets until the original chain is completely removed. This process can be reversed to refit the cleaned and lubricated original chain, thus removing the need to fully dismantle the chaincase and to remove the left-hand engine cover which would otherwise be necessary to feed the chain back on to the sprockets after cleaning.

3   Chain lubrication is most effectively accomplished by the use of a special chain grease such as Linklyfe or Chainguard. This is done by removing the chain as described above, cleaning it thoroughly in a petrol/paraffin mixture to remove all traces of old lubricant or road dirt, and immersing the chain in the grease which should be heated according to the manufacturer's instructions. This long and potentially messy process ensures that the innermost bearings in the chain are fully cleaned and lubricated. The grease itself is of a special type which stays on the chain more readily and is not flung off by centrifugal force as easily as thinner lubricants. A better solution for routine maintenance is the use of one of the many proprietary chain greases applied with an aerosol can. These are far easier to use, and very much cleaner, but should only be thought of as an addition to Linklyfe or Chainguard, and not a substitute for them. In spite of the manufacturer's claims, lubricant applied to the exterior of the chain by aerosol cannot penetrate to the

Tyre changing sequence — tubed tyres

 Deflate tyre. After pushing tyre beads away from rim flanges push tyre bead into well of rim at point opposite valve. Insert tyre lever adjacent to valve and work bead over edge of rim.

Use two levers to work bead over edge of rim. Note use of rim protectors.

 Remove inner tube from tyre.

When first bead is clear, remove tyre as shown.

 When fitting, partially inflate inner tube and insert in tyre.

Work first bead over rim and feed valve through hole in rim. Partially screw on retaining nut to hold valve in place.

 Check that inner tube is positioned correctly and work second bead over rim using tyre levers. Start at a point opposite valve.

Work final area of bead over rim whilst pushing valve inwards to ensure that inner tube is not trapped.

inner bearing surfaces of the chain as effectively as molten grease. Ordinary engine oil can be used, but only if nothing better is available, as it is far too thin, and too easily flung off, to lubricate the chain effectively.

4   Chain adjustment will need to be carried out at regular intervals to compensate for wear. Due to the fact that this wear never takes place evenly along the length of the chain, tight spots will develop which must be allowed for during adjustment.

5   Place the machine securely on its centre stand on level ground with the transmission in neutral. Remove the chaincase inspection cap. Rotate the rear wheel slowly with one hand, testing the chain tension with a finger at points along the entire length of the chain until the tightest spot is found. There should be 10 – 20 mm ($\frac{3}{8}$ – $\frac{3}{4}$ inch) total up and down movement, or free play, in the chain at this spot.

6   If there is more or less free play than the specified amount, remove the split pin securing the rear wheel spindle nut, and slacken the spindle nut and the large sleeve nut securing the sprocket assembly by just enough to permit the chain adjusters to move the spindle backwards or forwards as required. Tighten or slacken the two nuts on the chain adjusters to draw the spindle back to allow it to be moved forward as necessary.

7   Always adjust the drawbolts an equal amount in order to preserve wheel alignment. The fork ends are clearly marked with a series of vertical lines above or below the adjusters, to provide a simple, visual check. If desired, wheel alignment can be checked by running a plank of wood parallel to the machine, so that it touches the side of the rear tyre. If wheel alignment is correct, the plank will be equidistant from each side of the front wheel tyre, when tested on both sides of the rear wheel. It will not touch the front wheel tyre if this tyre is of smaller cross section. See the accompanying diagram.

8   On completion of chain adjustment, tighten the wheel spindle nuts to the torque figure given in the Specifications Section of this Chapter and fit a new split-pin. Check that the rear wheel rotates freely and check the rear brake pedal free play.

9   Do not run the chain overtight to compensate for uneven wear. A tight chain will place excessive stresses on the gearbox and rear wheel bearings, leading to their early failure. It will also absorb a surprising amount of power.

10   When there is no more room for adjustment the chain may be worn out, and in need of renewal. A simple check for this is to remove the lower chaincase half as described above and to try to pull the chain backwards off the rear sprocket. With the chain tensioned correctly, by application of hand pressure if this is no longer possible using the adjusters, it should not be possible to pull the chain clear of the sprocket teeth. If this can

be done, the chain must be considered worn out.

11   A more accurate assessment of chain condition involves removing it from the machine as described above. Wash the chain thoroughly in a petrol/paraffin mixture to remove all traces of old lubricant and to ensure that the maximum amount of free play is present. Lay the chain out lengthwise in a straight line and compress it endwise until all the free play is taken up. Measure the length of the chain, anchor one end and pull the chain out as far as possible. Measure the new length. If the chain has extended by more than $\frac{1}{4}$ inch per foot of original length it should be renewed.

12   Note that if the chain is to be renewed, this should only be done in conjunction with both sprockets. Running a new chain on worn sprockets will very rapidly wear the chain out.

13   When replacing the chain, make sure that the spring link is seated correctly, with the closed end facing the direction of travel.

14   An equivalent British-made chain of the correct size is available from Renold Limited. When ordering a new chain always quote the size (length and width at each pitch), the number of links and the machine to which it is fitted. For example, the standard chain is 428 (chain size) by 108 links. This will obviously vary if non-standard size sprockets are fitted.

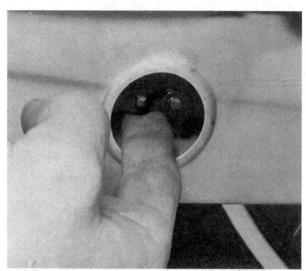

11.5 Find tightest spot in chain before measuring free play

11.6a Remove split pin and slacken wheel spindle nut ...

11.6b ... and then slacken large sprocket retaining nut –
H100

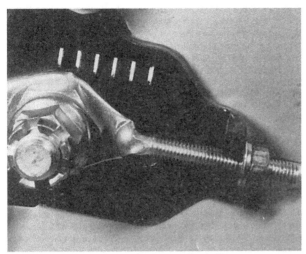

11.7 Use vertical marks to check rear wheel alignment

11.13a When replacing chain connecting link ...

11.13b ... always replace spring link with the closed end facing direction of travel

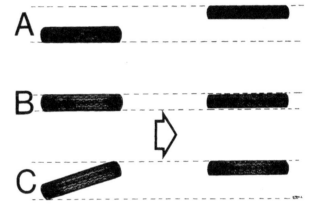

Fig. 5.5 Method of checking wheel alignment

A & C – Incorrect
B – Correct

12 Tyres: removal, repair and refitting

1   At some time or other the need will arise to remove and replace the tyres, either as a result of a puncture or because replacements are necessary to offset wear. To the inexperienced, tyre changing represents a formibable task, yet if a few simple rules are observed and the technique learned, the whole operation is surprisingly simple.

2   To remove the tyre from either wheel, first detach the wheel from the machine. Deflate the tyre by removing the valve core, and when the tyre is fully deflated, push the bead away from the wheel rim on both sides so that the bead enters the centre well of the rim. Remove the locking ring and push the tyre valve into a tyre itself.

3   Insert a tyre lever close to the valve and lever the edge of the tyre over the outside of the rim. Very little force should be necessary; if resistance is encountered it is probably due to the fact that the tyre beads have not entered the well of the rim, all the way round. If aluminium rims are fitted, damage to the soft alloy by tyre levers can be prevented by the use of plastic rim protectors.

4   Once the tyre has been edged over the wheel rim, it is easy to work round the wheel rim, so that the tyre is completely free

from one side. At this stage the inner tube can be removed.

5   Now working from the other side of the wheel, ease the other edge of the tyre over the outside of the wheel rim that is furthest away. Continue to work around the rim until the tyre is completely free from the rim.

6   If a puncture has necessitated the removal of the tyre, reinflate the inner tube and immerse it in a bowl of water to trace the source of the leak. Mark the position of the leak, and deflate the tube. Dry the tube, and clean the area around the puncture with a petrol soaked rag. When the surface has dried, apply rubber solution and allow this to dry before removing the backing from the patch, and applying the patch to the surface.

7   It is best to use a patch of self vulcanizing type, which will form a permanent repair. Note that it may be necessary to remove a protective covering from the top surface of the patch after it has sealed into position. Inner tubes made from a special synthetic rubber may require a special type of patch and adhesive, if a satisfactory bond is to be achieved.

8   Before replacing the tyre, check the inside to make sure that the article that caused the puncture is not still trapped inside the tyre. Check the outside of the tyre, particularly the tread area to make sure nothing is trapped that may cause a further puncture.

9   If the inner tube has been patched on a number of past

occasions, or if there is a tear or large hole, it is preferable to discard it and fit a replacement. Sudden deflation may cause an accident, particularly if it occurs with the rear wheel.

10 To replace the tyre, inflate the inner tube for it just to assume a circular shape but only to that amount, and then push the tube into the tyre so that it is enclosed completely. Lay the tyre on the wheel at an angle, and insert the valve through the rim tape and the hole in the wheel rim. Attach the locking ring on the first few threads, sufficient to hold the valve captive in its correct location.

11 Starting at the point furthest from the valve, push the tyre bead over the edge of the wheel rim until it is located in the central well. Continue to work around the tyre in this fashion until the whole of one side of the tyre is on the rim. It may be necessary to use a tyre lever during the final stages.

12 Make sure there is no pull on the tyre valve and again commencing with the area furthest from the valve, ease the other bead of the tyre over the edge of the rim. Finish with the area close to the valve, pushing the valve up into the tyre until the locking ring touches the rim. This will ensure that the inner tube is not trapped when the last section of bead is edged over the rim with a tyre lever.

13 Check that the inner tube is not trapped at any point. Reinflate the inner tube, and check that the tyre is seating correctly around the wheel rim. There should be a thin rib moulded around the wall of the tyre on both sides, which should be an equal distance from the wheel rim at all points. If the tyre is unevenly located on the rim, try bouncing the wheel when the tyre is at the recommended pressure. It is probable that one of the beads has not pulled clear of the centre well.

14 Always run the tyres at the recommended pressures and never under or over inflate. The correct pressures are given in the Specifications Section of this Chapter.

15 Tyre replacement is aided by dusting the side walls, particularly in the vicinity of the beads, with a liberal coating of french chalk. Washing up liquid can also be used to good effect, but this has the disadvantage, where steel rims are used, of causing the inner surface of the wheel rim to rust.

16 Never replace the inner tube and tyre without the rim tape in position. If this precaution is overlooked there is a good chance of the ends of the spoke nipples chafing the inner tube and causing a crop of punctures.

17 Never fit a tyre that has a damaged tread or sidewalls. Apart from legal aspects, there is a very great risk of a blowout, which can have very serious consequences on a two wheeled vehicle.

18 Tyre valves rarely give trouble, but it always advisable to check whether the valve itself is leaking before removing the tyre. Do not forget to fit the dust cap, which forms an effective extra seal.

Chapter 6 Electrical system

For specifications and information relating to the H100 S II model, refer to Chapter 7

Contents

Specifications

Battery
Make	Yuasa
Capacity	6V4 Ah
Earth connection	Negative (–)

Fuse 10A

Charging system
Generator output .... 74W @ 5000 rpm

Charge starts at:
Day .... 1500 rpm
Night .... 2000 rpm

	Day	Night
Charging rate – H100:		
Minimum @ 4000 rpm	1.7A/8.7V	1.3A/8.7V
Maximum @ 8000 rpm	3.5A/9.2V	3.0A/8.8V
Charging rate – H100 S:		
Minimum @ 4000 rpm	0.8A/8.8V	1.2A/8.7V
Maximum @ 8000 rpm	2.5A/8.9V	2.8A/8.9V

Bulbs
	H100	H100 S
Headlamp	6V 35/35W	6V 25/25W
Stop/tail lamp	6V 21/5W	6V 21/5W
Indicator lamp	6V 21W	6V 18W
Parking lamp	6V 4W	6V 4W
Speedometer and neutral indicator lamps	6V 3W	6V 3W
Indicator warning lamp	6V 1.7W	6V 1.7W

1 General description

The electrical system is powered by a flywheel generator fitted to the crankshaft left-hand end. The generator consists of a permanent magnet rotor and a multi-coil stator, only one of the coils on the stator being the power source for the lighting system, the other two are ignition system components.

The alternating current (ac) provided by the generator is fed largely to the lights, excess current being soaked up by a resistor when the main lighting switch is in the 'Off' or the 'P' position. The remainder of the generator output is converted to direct current (dc) by a silicon rectifier and is then used to charge the battery which powers the ancillary electrical equipment such as horn, stop lamp and flashing indicator lamps.

2 Testing the electrical system

1   Simple continuity checks, for instance when testing switch units, wiring and connections, can be carried out using a battery and bulb arrangement to provide a test circuit. For most tests described in this Chapter, however, a pocket multimeter should be considered essential. A basic multimeter capable ot measuring volts and ohms can be bought for a very reasonable sum and will prove an invaluable tool. Note that separate volt and ohmmeters may be used in place of the multimeter, provided those with the correct operating ranges are available. In addition, if the generator output is to be checked, an ammeter of 0-5 amperes range will be required.
2   Care must be taken when performing any electrical test, because some of the electrical components can be damaged if they are incorrectly connected or inadvertently shorted to earth. This is particularly so in the case of electronic components. Instructions regarding meter probe connections are given for each test, and these should be read carefully to preclude accidental damage occurring.
3   Where test equipment is not available, or the owner feels unsure of the procedure described, it is strongly recommended that professional assistance is sought. Errors made through carelessness or lack of experience can so easily lead to damage and need for expensive replacement parts.
4   A certain amount of preliminary dismantling will be necessary to gain access to the components to be tested. Normally, removal of the side panel will be required, with the possible addition of the fuel tank and headlamp unit to expose the remaining components.
5   Note that procedures for testing ignition system related components, including the main ignition switch, are given in the relevant Section of Chapter 3.

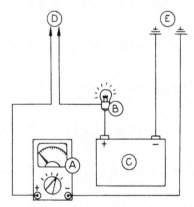

Fig. 6.1 Method of checking the wiring using a battery and bulb or a multimeter

A Multimeter   D Positive probe
B Bulb   E Negative probe
C Battery

3 Wiring: layout and examination

1   The wiring harness is colour-coded and will correspond with the accompanying wiring diagram. When socket connectors are used, they are designed so that reconnection can be made in the correct position only.

2   Visual inspection will usually show whether there are any breaks or frayed outer coverings which will give rise to short circuits. Occasionally a wire may become trapped between two components, breaking the inner core but leaving the more resilient outer cover intact. This can give rise to mysterious intermittent or total circuit failure. Another source of trouble may be the snap connectors and sockets, where the connector has not been pushed fully home in the outer housing, or where corrosion has occurred.
3   Intermittent short circuits can often be traced to a chafed wire that passes through or is close to a metal component such as a frame member. Avoid tight bends in the lead or situations where a lead can become trapped between casings.

4 Flywheel generator: checking the output

1   The generator output can only be checked with special equipment of the multimeter type. In the event of such equipment not being available, if the generator is suspect it should be checked by an authorised Honda dealer or by an auto-electrical specialist. Before any tests are carried out, the battery must be fully charged and the engine must be warmed up to normal operating temperature.
2   Remove the side panel to expose the battery and its leads. Connect a dc voltmeter (or a multimeter set to the relevant scale) across the battery by disconnecting the battery positive (+) lead, the red wire, and connecting one of the voltmeter leads to the positive (+) terminal lead. The other voltmeter lead is to be connected to the battery negative (−) lead, the blue wire. This should be done by pushing a metal probe inside the clear plastic insulation around the snap connector to make contact with the battery terminal lead without disconnecting it. An ammeter should now be connected between the battery positive (+) terminal lead and the positive (+) or red wire which forms part of the main loom.
3   Start the engine and note the readings taken at the engine speeds given in the specifications Section of this Chapter. Compare your readings with those given. Switch on the lights at the main lighting switch and turn on main beam. Take another, similar, set of readings which should be compared with those given for night time charging rates.
4   If the readings obtained are correct, it may be assumed that the charging system is functioning properly. A marked reduction in output caused by a generator fault may be a result of damaged windings in the stator coil or damaged leads. These may be checked for continuity and resistance by a relatively simple test which does not require removal of the generator. Tests for the other components in the charging system are given in the following Sections.
5   Disconnect the voltmeter and ammeter and connect the battery up again. Trace the generator lead from the engine up to the snap connectors on the right-hand side of the machine. Identify and disconnect the yellow, green (where applicable), and white wires. Using a multimeter set to the resistance mode, check for resistance between the wires specified and compare the readings obtained with those given:

Yellow to green (or earth)   0.1 − 1.0 ohm
White to green (or earth)   0.3 − 1.5 ohm

6   If the readings obtained are not satisfactory, take the machine to an authorised Honda dealer for checking as the only practical solution to a fault in the stator is the renewal of the generator assembly as a complete unit. The stator and rotor are not available as separate items, neither are any individual components. This course of action is, however, likely to prove expensive and so an expert second opinion is advisable. If, on closer inspection, a wire turns out to be damaged or broken, repair is relatively easy for the expert. The payment of a small sum for the time involved in such a repair is infinitely preferable to paying a large amount of money for unnecessary new parts.

7   If the first test indicates a fault exists in the charging system but the generator is proven to be in good order by the second test, the next component in the charging system to be tested is the silicon rectifier unit.

5   Rectifier: location and testing

1   The silicon rectifier is a small rectangular plastic block with two male spade terminals which fit a plastic two pin connector block from the wiring loom. It is situated behind the right-hand side panel in a holder immediately in front of the battery, on H100 models, or on the frame top tube on H100 S models.

2   The rectifier consists of a small diode and serves to convert the ac output of the flywheel generator into dc to charge the battery. It should be thought of as a one-way valve, in that it will allow the current to flow in one direction only, thus blocking half of the output wave from the generator.

3   Before removing the unit, identify the polarity of the two terminals by the colour of the wire leading to each one. The red wire leads to the positive (+) terminal and the white wire to the negative (–) terminal.

4   Using a multimeter set to the resistance mode, check for continuity between the two terminals. There should only be continuity from the negative (–) to the positive (+) terminal. This direction of flow may, depending on the make of the rectifier fitted, be shown by an arrow marking moulded into the top surface of the unit. If there is continuity in the reverse direction, or if resistance is measured in both directions, the rectifier is faulty and must be renewed. No repair is possible.

5.1b Location of rectifier – H100 S

6.1 Location of resistor (arrowed)

5.1a Location of rectifier (arrowed) – note also spare fuse and fuse holder – H100

6   Resistor: location and testing

1   This unit is an oblong metal component with a single pink wire leading to it. It is bolted to the underside of the frame on H100 models, midway between the steering head and the engine top mounting. While it can be reached from underneath, access is greatly improved if the petrol/oil tank is first removed as described in Section 2 of Chapter 2. On H100 S models it is bolted to the bottom yoke.

2   The resistor's function is to soak up the unused current produced by the generator when the main lighting switch is in the 'Off' or the 'P' position. It therefore serves to protect bulbs and wiring from overcharging. If bulb blowing or other symptoms associated with overcharging are encountered it may be at fault. The test procedure is extremely simple.

3   Disconnect the pink wire at its snap connector, and using a multimeter set to the resistance function, measure the amount

of resistance between the terminal of the pink wire and a suitable earth point on the frame. The set figure is 1 ohm on H100 models, 2 ohm on H100 S models. If the measured figure is appreciably higher than this, carefully check that the resistor is properly earthed at its mounting point, and that the mating surfaces are clean and free from paint, dirt, and corrosion. Once it is known that the earth connections are in good order, repeat the test. If the figure is still too high, or if it was too low in the first place, the resistor must be renewed.

7   Battery: examination and maintenance

1   The battery is located behind the right-hand side panel (left-hand, H100 S). It is held by a rubber strap in a tray formed by the back of the air filter housing. To remove it, withdraw the side panel which is retained by a single screw at the front and by a clip at the rear, disconnect the two battery leads at their snap connectors and release the rubber retaining strap. Withdraw the battery complete with its vent tube as this latter should be inspected regularly to ensure that it remains free from blockages. Reverse the above procedure to refit the battery.

2   The transparent plastic case of the battery permits the upper and lower levels of the electrolyte to be observed, without disturbing the battery, by removing the side cover. Maintenance is normally limited to keeping the electrolyte level between the

prescribed upper and lower limits and making sure that the vent tube is not blocked. The lead plates and their separators are also visible through the transparent case, a further guide to the general condition of the battery. If electrolyte level drops rapidly suspect over-charging and check the system.

3   Unless acid is spilt, as may occur if the machine falls over, the electrolyte should always be topped up with distilled water to restore the correct level. If acid is spilt onto any part of the machine, it should be neutralised with an alkali such as washing soda or baking powder and washed away with plenty of water, otherwise serious corrosion will occur. Top up with sulphuric acid of the correct specific gravity (1.260 to 1.280) only when spillage has occurred. Check that the vent pipe is well clear of the frame or any of the other cycle parts.

4   It is seldom practicable to repair a cracked battery case because the acid present in the joint will prevent the formation of an effective seal. It is always best to renew a cracked battery, especially in view of the corrosion which will be caused if the acid continues to leak.

5   If the machine is not used for a period of time, it is advisable to remove the battery and give it a 'refresher' charge every six weeks or so from a battery charger. The battery will require recharging when the specific gravity falls below 1.260 (at 29°C – 68°F). The hydrometer reading should be taken at the top of the meniscus with the hydrometer vertical. If the battery is left discharged for too long, the plates will sulphate. This is a grey deposit which will appear on the surface of the plates, and will inhibit recharging. If there is sediment on the bottom of the battery case, which touches the plates, the battery needs to be renewed. Prior to charging the battery refer to the following Section for correct charging rate and procedure. If charging from an external source with the battery on the machine, disconnect the leads, or the rectifier will be damaged.

6   Note that when moving or charging the battery, it is essential that the following basic safety precautions are taken:

a)   Before charging check that the battery vent is clear or, where no vent is fitted, remove the combined vent/filler caps. If this precaution is not taken the gas pressure generated during charging may be sufficient to burst the battery case, with disastrous consequences.

b)   Never expose a battery on charge to naked flames or sparks. The gas given off by the battery is highly explosive.

c)   If charging the battery in an enclosed area, ensure that the area is well ventilated.

d)   Always take great care to protect yourself against accidental spillage of the sulphuric acid contained within the battery. Eyeshields should be worn at all times. If the eyes become contaminated with acid they must be flushed with fresh water immediately and examined by a doctor as soon as possible. Similar attention should be given to a spillage of acid on the skin.

Note also that although, should an emergency arise, it is possible to charge the battery at a more rapid rate than that stated in the following Section, this will shorten the life of the battery and should therefore be avoided if at all possible.

8   Battery: charging procedure

1   Whilst the machine is used on the road it is unlikely that the battery will require attention other than routine maintenance because the generator will keep it fully charged. However, if the machine is used for a succession of short journeys only, mainly during the hours of darkness when the lights are in full use, it is possible that the output from the generator may fail to keep pace with the heavy electrical demand, especially if the machine is parked with the lights switched on. Under these circum-

7.1 Right-hand side panel is retained by a single screw at the front – H100

7.2 Transparent casing simplifies battery inspection

stances it will be necessary to remove the battery from time to time to have it charged independently.

2   The normal maximum charging rate for any battery is 1/10 the rated capacity. Hence the charging rate for the 4 Ah battery fitted to this machine is 0.4 amp. A slightly higher charge rate may be used in emergencies only, but this should not exceed 1 amp.

3   Ensure that the battery/charger connections are properly made, ie the charger positive (usually coloured red) lead to the battery positive (the red wire) lead, and the charger negative (usually coloured black or blue) lead to the battery negative (the blue wire) lead. Refer to the previous Section for precautions to be taken during charging. It is especially important that the battery cell cover plugs are removed to eliminate any possibility of pressure building up in the battery and cracking its casing. Switch off the charger if the cells become overheated, ie over 45°C (117°F).

4   Charging is complete when the specific gravity of the electrolyte rises to 1.260 – 1.280 at 20°C (68°F). A rough guide to this state is when all three cells are gassing freely. At the normal (slow) rate of charge this will take between 3 – 15 hours, depending on the original state of charge of the battery.

5   If the higher rate of charge is used, never leave the battery charging for more than 1 hour as overheating and buckling of the plates will inevitably occur.

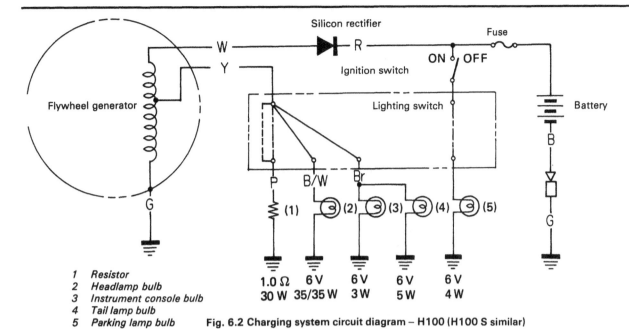

1　Resistor
2　Headlamp bulb
3　Instrument console bulb
4　Tail lamp bulb
5　Parking lamp bulb

1.0 Ω	6 V	6 V	6 V	6 V
30 W	35/35 W	3 W	5 W	4 W

Fig. 6.2 Charging system circuit diagram – H100 (H100 S similar)

9　Fuse: location and renewal

1　The electrical system is protected by a single fuse of 10 amp rating. It is retained in a plastic casing set in the battery positive (+) terminal lead, and is clipped to a holder immediately in front of the battery. A spare fuse is clipped alongside the main fuse holder. If the spare fuse is ever used, replace it with one of the correct rating as soon as possible.

2　Before renewing a fuse that has blown, check that no obvious short circuit has occurred, otherwise the replacement fuse will blow immediately it is inserted. It is always wise to check the electrical circuit thoroughly, to trace the fault and eliminate it.

3　When a fuse blows while the machine is running and no spare is available, a 'get you home' remedy is to remove the blown fuse and wrap it in silver paper before replacing it in the fuse holder. The silver paper will restore the electrical continuity by bridging the broken fuse wire. This expedient should never be used if there is evidence of short circuit or other major electrical fault, otherwise more serious damage will be caused. Replace the 'doctored' fuse at the earliest possible opportunity, to restore full circuit protection.

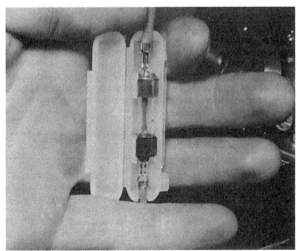

9.1 Fuse is retained in plastic casing

10　Horn: location and testing

1　The horn is mounted on the bottom yoke via a flexible metal strip retained by a single bolt. No maintenance is required, or indeed possible, other than regular cleaning to remove road dirt and occasional spraying with WD40 or a similar water dispersant spray to minimise internal corrosion.

2　If the horn fails to work, first check that the battery is fully charged. If full power is available, a simple test will reveal whether the current is reaching the horn. Disconnect the horn wires and substitute a 6 volt bulb. Switch on the ignition and press the horn button. If the bulb fails to light, check the horn button and wiring as described elsewhere in this Chapter. If the bulb does light, the horn circuit is proved good and the horn itself must be checked.

3　With the horn wires still disconnected, connect a fully charged 6 volt battery directly to the horn. If it does not sound, a sharp tap on the outside may serve to free the internal contacts. If this fails, the horn must be renewed as repair and adjustment are not possible.

10.1 Horn is mounted by a flexible strip on the bottom yoke

11 Flashing indicator relay: location and testing

1   The flashing indicator relay is situated underneath the petrol/oil tank, clipped to the frame by a flexible rubber mounting. It is a very delicate unit which is rubber mounted to protect it from vibration. Handle it carefully at all times. No maintenance is necessary, and if it is found to be faulty, renewal is the only practical solution.

2   If the flashing indicator lamps cease to function correctly, there may be any one of several possible faults responsible which should be checked before the relay is suspected. First check that the flashing indicator lamps are correctly mounted and that all the earth connections are clean and tight. Check that the bulbs are of the correct wattage and that corrosion has not developed on the bulbs or in their holders. Any such corrosion must be thoroughly cleaned off to ensure proper bulb contact. Also check that the flashing indicator switch is functioning correctly and that the wiring is in good order. Finally ensure that the battery is fully charged.

3   Faults in any one or more of the above items will produce symptoms for which the flashing indicator relay may be blamed unfairly. If the fault persists even after the preliminary checks have been made, the relay must be at fault. Unfortunately the only practical method of testing the relay is to substitute a known good one. If the fault is then cured, the relay is proven faulty and must be renewed. Fortunately relay failure is a rare occurrence.

4   If renewal is necessary, the petrol/oil tank must be removed first, as described in Section 2 of Chapter 2. Disconnect the two wires, noting the terminal to which each wire is attached and withdraw the relay. On refitting, some replacement Honda relays have the two terminals marked with correctly-coloured blobs of paint to indicate which wire goes to which terminal. If these marks are not present, replace the wires in exactly the same way as on the original unit.

12 Bulb renewal: headlamp

1   To renew the headlamp bulb, slacken and remove the screw(s) which retains the headlamp rim to the headlamp casing and withdraw the headlamp rim/reflector unit assembly. The bulb holder is retained in the headlamp reflector by a clip on its upper edge and by a coil spring at its lower edge. Lift the bulb holder up against spring pressure and disengage its mounting clip from the reflector unit. Depress the bulb in the holder, turn it gently anti-clockwise and remove it. Reassembly is a straightforward reversal of this procedure. On H100 S models, press the bulbholder in and twist to release it, then remove the bulb.

2   The parking lamp bulb holder is a push fit in a rubber grommet set in the reflector unit. Withdraw the bulb holder and remove the bulb in the same way as the headlamp bulb.

3   Vertical headlamp beam adjustment is made by slackening the two headlamp mounting bolts and tilting the headlamp assembly as necessary. Note that reference marks on the headlamp casing must be aligned with index marks stamped in the headlamp brackets. If this is done on reassembly it will serve as a basis for proper beam adjustment.

4   In the UK, regulations stipulate that the headlamps must be arranged so that the light will not dazzle a person standing at a distance greater than 25 feet from the lamp, whose eye level is not less than 3 feet 6 inches above that plane. It is easy to approximate this setting by placing the machine 25 feet away from a wall, on a level road, and setting the dipped beam height so that it is concentrated at the same height as the distance of the centre of the headlamp from the ground. The rider must be seated normally during this operation, and also the pillion passenger, if one is carried regularly.

11.1a Location of flashing indicator relay – H100

11.1b Location of flashing indicator relay – H100 S

12.1a To remove headlamp bulb unscrew headlamp rim mounting screw and withdraw headlamp/reflector unit assembly

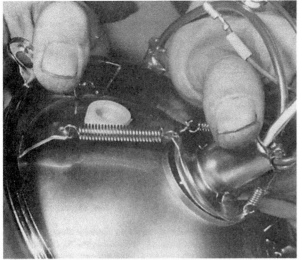

12.1b Lift bulb holder up against spring tension – H100 ...

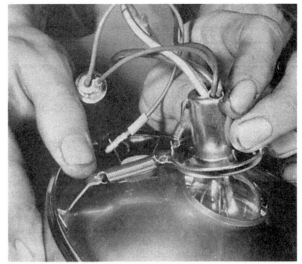

12.1c ... and disengage mounting clip

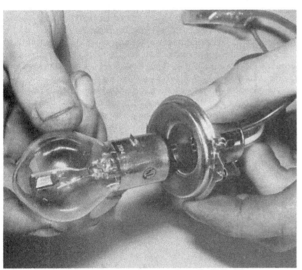

12.1d Gently twist bulb to disengage from bulb holder – H100

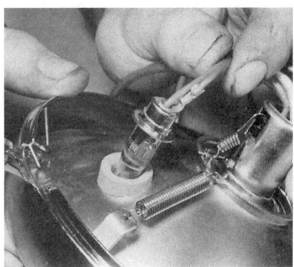

12.2 Parking lamp bulb is rubber mounted in reflector unit

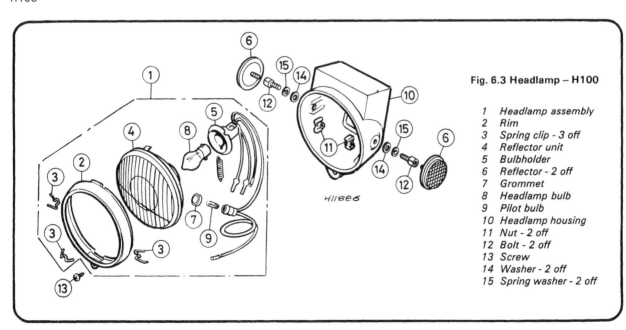

Fig. 6.3 Headlamp – H100

1   Headlamp assembly
2   Rim
3   Spring clip - 3 off
4   Reflector unit
5   Bulbholder
6   Reflector - 2 off
7   Grommet
8   Headlamp bulb
9   Pilot bulb
10 Headlamp housing
11 Nut - 2 off
12 Bolt - 2 off
13 Screw
14 Washer - 2 off
15 Spring washer - 2 off

H11826

13 Bulb renewal: stop and tail lamp

1   The combined stop and tail lamp bulb contains two filaments, one for the stop lamp and one for the tail lamp.
2   The offset pin bayonet fixing bulb can be removed after the plastic lens cover and screws have been removed.

14 Bulb renewal: flashing indicator lamps

1   The flashing indicator lamp assemblies are plastic mouldings bolted to metal stalks. The earth connections should be checked at convenient intervals to ensure that they are tight and free from dirt or corrosion. If a bulb or wiring connection fails, the affected lamp will cease operation, the failure being indicated by rapid flashing of the remaining bulb.
2   The lens is clipped to the body of the lamp and can be removed by using a coin or broad-bladed screwdriver in the slot provided to lever the lens off. Both the lens and the body are of plastic construction, so care must be taken to avoid damage during removal. The bulb is of the conventional bayonet type and can be removed by pushing inwards, twisting gently anti-clockwise and releasing.
3   No reflectors are fitted to the indicator lamps, and owners may wish to improve their visibility in bright sunlight by lining the inside of the lamp body with aluminium cooking foil. The lamps are, however, adequate in most normal conditions.

15 Bulb renewal: instrument panel and warning lamps

1   The various bulbs in the instrument panel are held in rubber holders which are a push fit in the underside of the panel. To gain access to the holders it is best to release the panel so that it can be tilted upwards. On H100 S models, remove the bottom cover.

2   The bulbs are of the bayonet cap type and are released by depressing and twisting them anti-clockwise on H100 models. On H100 S models they are of the capless type and are pulled out of their connectors. When purchasing replacement bulbs ensure that they are of the specified voltage and wattage.

16 Switches: general

1   Generally speaking, the switches fitted to the machines covered in this Manual should give little trouble, but if necessary they may be tested as described in the following Sections.
2   To test the switches, a multimeter, set to the resistance function, will be required to carry out the various continuity checks described.
3   On models fitted with a battery, always disconnect the battery before removing any of the switches, to prevent the possibility of a short circuit. Most troubles are caused by dirty contacts, but in the event of the breakage of some internal part, it will be necessary to renew the complete switch.
4   It should, however, be noted that if a switch is tested and found to be faulty, there is nothing to be lost by attempting a repair. It may be that worn contacts can be built up with solder, or that a broken wire terminal can be repaired, again using a soldering iron. The handlebar switches can all be dismantled to a greater or lesser extent, but the ignition and stop lamp switches are sealed units. It is, however, up to the owner to decide if he has the skill to carry out this sort of work.
5   While none of the switches require routine maintenance of any sort, some regular attention will prolong their life a great deal. In the author's experience, the regular and constant application of WD40 or a similar water-dispersant spray not only prevents problems occurring due to waterlogged switches and the resulting corrosion, but also makes the switches much easier and more positive to use. Alternatively, the switch may be packed with a silicone-based grease to achieve the same result.

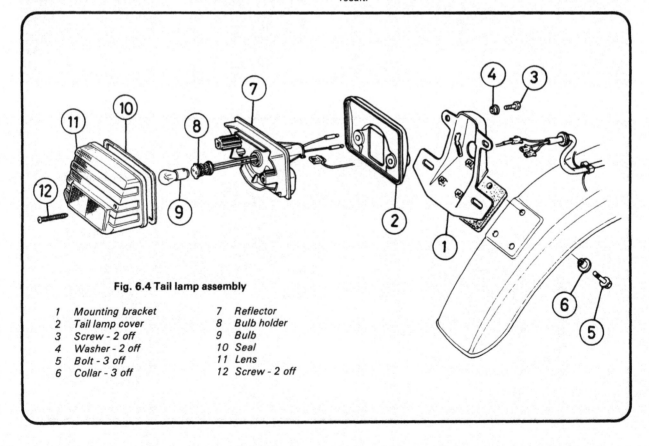

Fig. 6.4 Tail lamp assembly

1 Mounting bracket	7 Reflector
2 Tail lamp cover	8 Bulb holder
3 Screw - 2 off	9 Bulb
4 Washer - 2 off	10 Seal
5 Bolt - 3 off	11 Lens
6 Collar - 3 off	12 Screw - 2 off

13.2a Release two screws to remove tail lamp lens

13.2b Bulb is a conventional bayonet fitting

14.2a Lens is levered away from lamp body

14.2b Twist bulb to disengage from bulb holder

15.1 Instrument bulb holders are a push fit in the underside of the panel

16.1 Handlebar switches generally give little trouble

17 Stop lamp switches: testing and adjustment

1   The front stop lamp switch is incorporated in the front brake
lever assembly and is automatically operated by the application
of the front brake. No adjustment is possible.
2   The rear stop lamp switch is located by a bracket on the
lower right-hand side of the frame and is connected to the rear
brake pedal by a spring. It should be adjusted to operate just as
the brake pedal has taken up its free play and is beginning to
engage the rear brake. This is achieved by turning the plastic
sleeve nut as required to raise or lower the switch.
3   Testing is an extremely simple process for either switch.
When testing the front switch, remove the headlamp and
identify the switch lead. Disconnect the two switch wires which
are coloured black and green/yellow respectively. Check for
continuity between these two wires when the front brake lever
is firmly applied. For the rear switch, simply remove the two
wires from the terminals on the top of the switch body and
check for continuity between the two terminals when the rear
brake pedal is pressed down.
4   If no continuity exists when the brake is applied, the switch
in question must be renewed. It is not possible to effect a repair
to either assembly.

18 Flashing indicator switch: testing

1   The indicator switch is a horizontally-mounted three-
position rocker switch. To test it, remove the headlamp, then
identify and disconnect the light blue, orange, and grey wires
from the main lead of the left-hand handlebar switch cluster.
2   Once this has been done, turn the switch to the 'R' position
and check for continuity between the light blue and grey wire
terminals. Turn the switch to the 'L' position and check for
continuity between the orange and grey wires. If continuity is
found on both sets of wires, the switch is in good order, but it
is worthwhile to check for continuity between all three wires
with the switch in the central or 'Off' position. There should be
no continuity at all in this position.
3   If the tests prove the switch to be faulty in any way it
should be renewed. Bear in mind, however, the observations
made in Section 16 concerning repairs.

19 Lighting switch: testing

1   The main lighting switch is a sliding three position unit
mounted vertically on the inner side of the left-hand handlebar
switch cluster. To test it, remove the headlamp as previously
described and disconnect the switch lead at the snap connectors. Also slacken and remove the two screws fastening both
halves of the switch cluster and separate the two switch halves.
Identify the wires as appropriate and the main feed between the
lighting switch and the dipswitch. Test for continuity between
the terminals in the following order.
2   With the switch in the 'Off' position, check the yellow and
pink wires. With the switch in the 'P' position, check the yellow,
pink, and brown together, and the black and brown/white
together. With the switch in the 'H' position, check between the
yellow and brown wires and the dipswitch feed together.
3   If continuity is found between all the wires in each group
when the switch is in the position indicated, the switch is in
good order. If not, the switch is faulty and should therefore be

renewed unless the owner considers himself competent to carry
out repairs. Remember that if the switch were to fail, causing a
total loss of lighting while riding at night, the consequences
might be most unpleasant. In the case of this switch it might be
better to err on the side of safety and to purchase a new switch
cluster.

20 Dipswitch: testing

1   The dipswitch is a sliding two-position switch mounted
vertically on the outer side of the left-hand handlebar switch
cluster. To test it, remove the headlamp as previously described
and identify, then disconnect, the blue and white wires from the
main lead of the left-hand handlebar switch cluster. Slacken and
remove the two screws fastening the two halves of the switch
cluster and separate the two switch halves. Identify the main
feed between the lighting switch and the dipswitch, and ensure
that the main lighting switch is in the 'H' position.
2   With the dipswitch in the 'Hi' position, check for continuity
between the blue wire and the main feed from the lighting
switch. Move the switch to the 'Lo' position and check for
continuity between the white wire and the main feed. Check
also that with the switch in the middle of its travel, there is
continuity between all three terminals.
3   If continuity exists in all cases the switch is serviceable. If
not, the only practical solution is the renewal of the switch
cluster. While repairs may be possible, and nothing is to be lost
by attempting repairs, the warning given at the end of the
previous Section should be borne in mind.

21 Horn button: testing

1   The horn button is a simple spring-loaded assembly on the
lower part of the left-hand handlebar switch cluster. A simple
test sequence is given in Section 10 of this Chapter.
2   If the horn button is proved faulty by this test it may prove
possible, depending on the owner's ability, to strip the switch
and repair the damage. If not, renewal of the switch cluster is
the only answer. It should be remembered that a horn in good
working order is a legal requirement in the UK.

22 Neutral indicator switch: testing

1   To test the neutral indicator switch, trace the generator lead
back up to the snap connectors on the right-hand side of the
machine in the vicinity of the battery. Identify and disconnect
the light green/red wire and the dark green. Ensure that the
transmission is in the neutral position and check for continuity
between the two wire terminals.
2   If continuity exists, the switch and wiring are serviceable,
and the fault is in the wiring between the function of the
generator lead to the main loom and the neutral indicator bulb.
If the bulb is known to be in working order the fault must be
traced and repaired. If continuity does not exist, remove the
engine left-hand outer cover and, with the transmission still in
neutral, check for continuity between the brass switch terminal
and the crankcase. If continuity is then found, the switch is
proven to be in good order and the fault is in the wiring of the
generator lead. If no continuity exists, the switch should be
renewed. Repairs are not likely to be possible.

The H100 S-G (H100 S II) model

Chapter 7 The H100 S II model

Contents

Specifications

Note: *The specifications shown below relate to the H100 S II model sold in the UK from March 1986 and covered in this update Chapter. Where no specifications are shown here, refer to those given for the H100 S model in the Specifications section at the beginning of each Chapter.*

Dimensions and weight
Overall length	1840 mm (72.4 in)
Overall width	690 mm (27.2 in)
Overall height	1020 mm (40.2 in)
Wheelbase	1200 mm (47.2 in)
Ground clearance	160 mm (6.3 in)
Dry weight	86 kg (189.5 lb)

Specifications relating to Chapter 1

Engine
Port timing:
Intake	Timing controlled by reed valve
Exhaust opens at	85° BBDC
Exhaust closes at	85° ABDC
Scavenge opens at	59° BBDC
Scavenge closes at	59° ABDC
Booster port opens at	61° BBDC
Booster port closes at	61° ABDC

Specifications relating to Chapter 3

Ignition system
Type	Capacitor discharge ignition (CDI)

Ignition high tension coil
Winding resistances:
Primary	0.16 – 0.20 Ω (Black terminal to green terminal)
Secondary	3.7 – 4.5 k Ω (HT lead to green terminal)

Specifications relating to Chapter 5

Tyres
Rear	2.75-18-4PR

Specifications relating to Chapter 6

Charging system
Generator output	79W ◉ 5000 rpm

Bulbs
Indicator lamp	6V 10W

1 Introduction

This update Chapter covers the H100 S II (or H100 S-G) model sold in the UK from March 1986 onward. This Mk II version of the H100 S is a largely cosmetic reworking of the previous H100 S-D model. The main visible change is the adoption of bolted-on frame downtubes, giving the machine the appearance of a conventional cradle-frame motorcycle. For all practical purposes these tubes are cosmetic; the main frame being similar to that of the previous H100 S model. Further changes include revised paint colours and graphics.

The H100 S II remains largely unchanged mechanically, although the ignition and electrical systems have been subject to revision. Most significantly, the contact breaker ignition which was fitted on the earlier H100 S-D, has been dropped in favour of capacitor discharge ignition (CDI) of the same type as used on the H100 A.

From January 1988 a revised H100 S II was introduced, the H100 S-J. Model changes are of a cosmetic nature.

2 CDI ignition system: general

1   As has been mentioned above, the H100 S II model employs a capacitor discharge ignition (CDI) system similar to that used on the H100 A machines. The use of this system dispenses with the contact breaker assembly, and thus the need for regular checking and adjustment.

2   For all practical purposes the system can be dealt with in the same way as described for the H100 A in Chapter 3. Note however that the ignition coil resistances are changed on the H100 S II, and these will be found in the specifications at the beginning of this Chapter. The ignition circuit is shown in the circuit diagram which accompanies this Section.

3 Ignition coil: testing

1   The ignition coil can be tested using the procedure described in Section 5 of Chapter 3. The primary and secondary winding resistances applicable to the H100 S II are given at the beginning of this Chapter, which differ from those of the previous models.

2   The resistance tests described will normally give a useful indication of coil condition. Note, however, that faults caused by the breakdown of insulation under load will not be revealed by these tests, for which a coil tester or CDI tester unit will be required. Where this type of fault is suspected, either fit a sound coil and note whether this resolves the problem, or take the coil to an authorized Honda dealer for testing.

4 CDI unit: testing

1   The CDI unit used on the H100 S II is identical to that fitted to the H100 A model. For details of the test procedures required, refer to Chapter 3, Section 6, and to Fig. 3.4. It should be noted that the resistance test may not reveal all potential faults, and that a full check requires an operational test to be performed using a CDI tester unit.

2   This equipment is expensive, and is likely to be available only at the larger Honda dealers. In the event of an elusive CDI fault, and in the absence of this equipment, it is preferable to eliminate the CDI unit as a likely cause of the problem by temporary substitution of a new unit.

5 Flywheel generator: testing the source and pulser coils

Like the H100 A, the H100 S II uses source and pulser coils within the flywheel generator to power the CDI system and to provide the trigger pulse which controls the ignition spark. For full details of the test procedures for these two sets of windings, refer to Chapter 3, Section 7.

6 Frame downtube assembly: general

1   As has been mentioned above, the H100 S II model is fitted with bolted-on frame downtubes to give the machine the

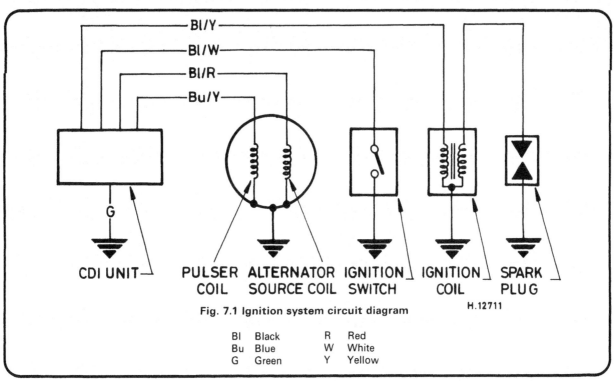

Fig. 7.1 Ignition system circuit diagram

H.12711

Bl	Black	R	Red
Bu	Blue	W	White
G	Green	Y	Yellow

appearance of a conventional cradle-frame motorcycle. Despite appearances, the machine remains unchanged in the sense that the engine/transmission unit hangs from the frame as before.

2   For most practical purposes, the downtubes will have little or no effect on working procedures. On occasions, however, it may prove advantageous to remove them to permit easier access, or become unavoidable if the engine/transmission unit is to be removed from the frame, or if the cylinder head and barrel need to be lifted. The assembly is attached to the frame at the top by a single bolt, and to the crankcase by a long through bolt. Two short bolts secure the assembly to the footrest bar.

7   Regulator/rectifier unit: location and testing

1   The electrical system on the H100 S II has been modified, a combined regulator/rectifier unit being fitted in place of the separate diode block (rectifier) and resistor used on the previous models. The regulator/rectifier is a finned alloy unit located just forward of the battery.

2   In the event of a charging system fault, the internal resistances of the unit should be checked using either a SANWA SP-10D or a KOWA TH-5H tester (multimeter). Honda caution that the internal circuitry of the regulator/rectifier unit can produce different readings on other meters. Some indication of the unit's condition can, however, be obtained, even if the exact resistance shown appears incorrect.

3   Using the accompanying illustration to identify the unit terminals, check the resistances with the meter probes connected as per the table. If any one reading is significantly different from that specified, it is likely that there is an internal

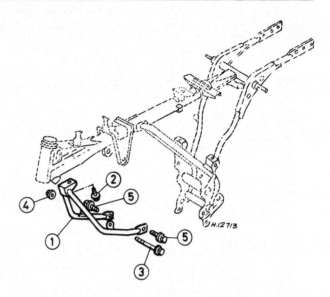

Fig. 7.2 Frame downtube mountings

1	Downtubes	4	Nut
2	Upper mounting bolt	5	Footrest bar mounting
3	Long through bolt		bolt – 2 off

fault in the unit. Should this be the case, it will be necessary to fit a new unit; its sealed construction rules out any attempt at repair.

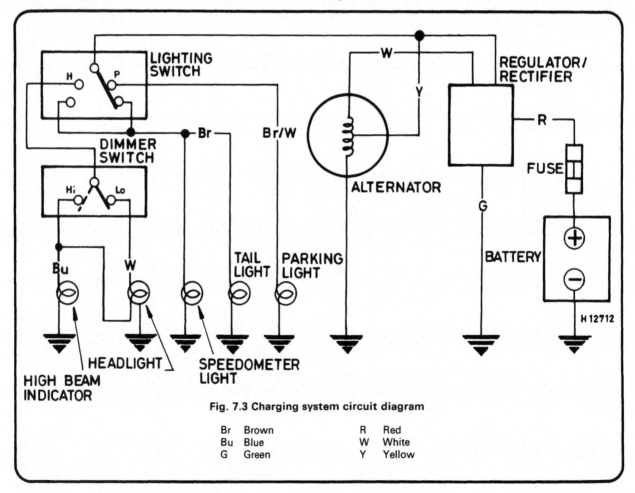

Fig. 7.3 Charging system circuit diagram

Br	Brown	R	Red
Bu	Blue	W	White
G	Green	Y	Yellow

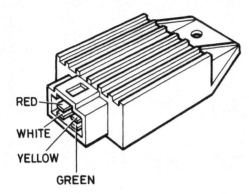

RED

WHITE

YELLOW

GREEN

Unit : kΩ

PROBE ⊖ ⊕ PROBE	White	Red	Yellow	Green
White		0.5-10	∞	∞
Red	∞		∞	∞
Yellow	∞	∞		10-500
Green	∞	∞	10-500	

H.12635

Fig. 7.4 Regulator/rectifier unit test

Sanwa tester – set to x KΩ scale
Kowa tester – set to x 100Ω scale

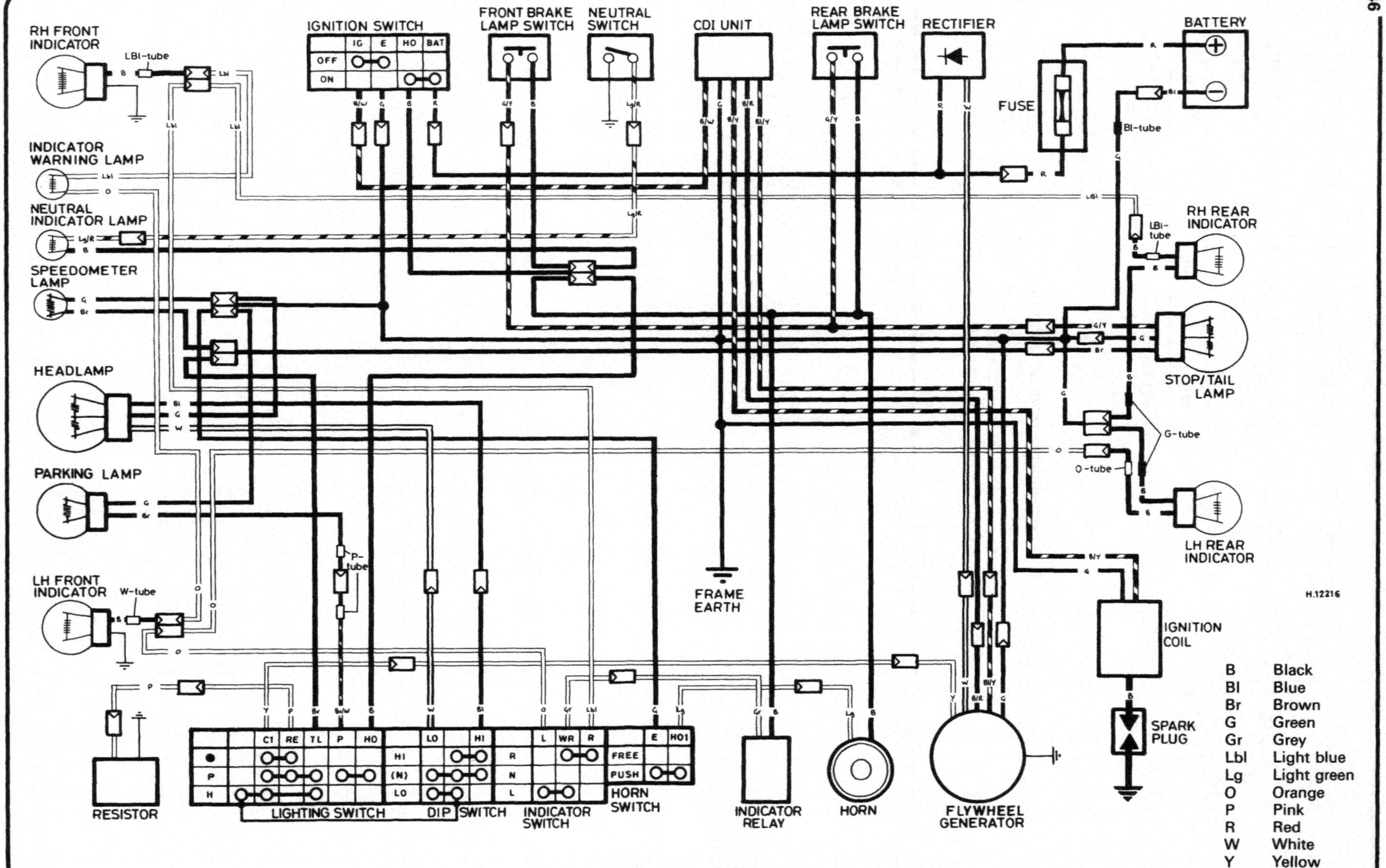

H.12216

Colour key:

B	Black
Bl	Blue
Br	Brown
G	Green
Gr	Grey
Lbl	Light blue
Lg	Light green
O	Orange
P	Pink
R	Red
W	White
Y	Yellow

BATTERY

RH REAR INDICATOR

STOP/TAIL LAMP

G-tube

LH REAR INDICATOR

IGNITION COIL

SPARK PLUG

Bl-tube

Lbl-tube

FUSE

RECTIFIER

O-tube

FLYWHEEL GENERATOR

REAR BRAKE LAMP SWITCH

HORN

CDI UNIT

FRAME EARTH

INDICATOR RELAY

NEUTRAL SWITCH

FRONT BRAKE LAMP SWITCH

IGNITION SWITCH

	IG	E	HO	BAT
OFF				
ON				

HORN SWITCH

E	HO1
FREE	
PUSH	

INDICATOR SWITCH

| L | WR | R |

DIP SWITCH

HI		
LO		
HI	(N)	LO

LIGHTING SWITCH

HO		
P		
TL		
C1	RE	TL

P-tube

W-tube

RH FRONT INDICATOR

Lbl-tube

INDICATOR WARNING LAMP

NEUTRAL INDICATOR LAMP

SPEEDOMETER LAMP

HEADLAMP

PARKING LAMP

LH FRONT INDICATOR

RESISTOR

Wiring diagram – H100-A model

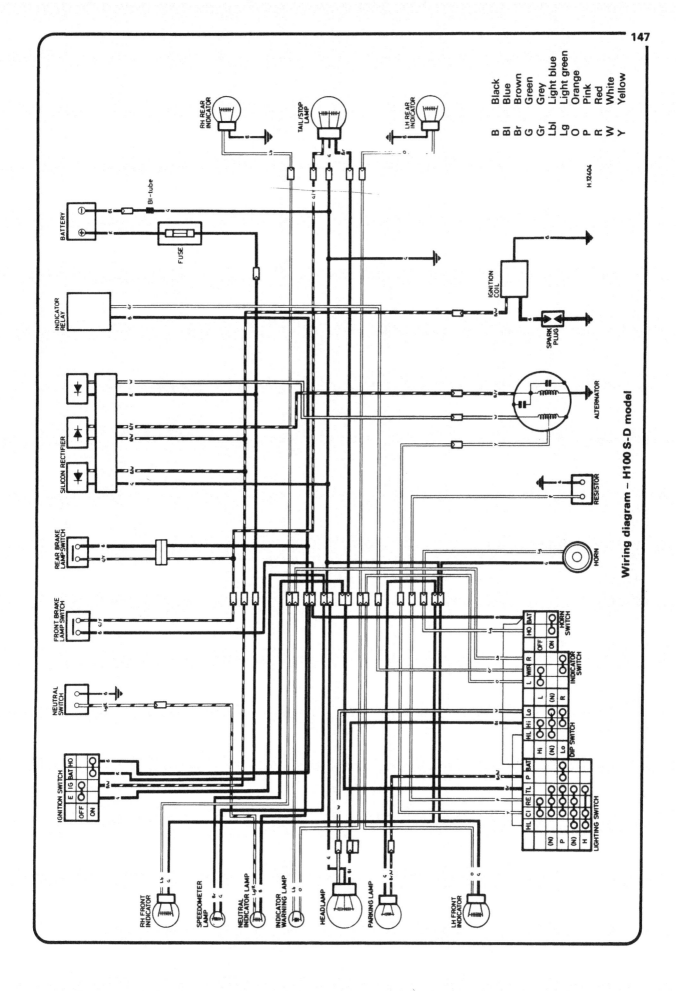

Wiring diagram – H100 S-D model

H.12404

B	Black
Bl	Blue
Br	Brown
G	Green
Gr	Grey
Lbl	Light blue
Lg	Light green
O	Orange
P	Pink
R	Red
W	White
Y	Yellow

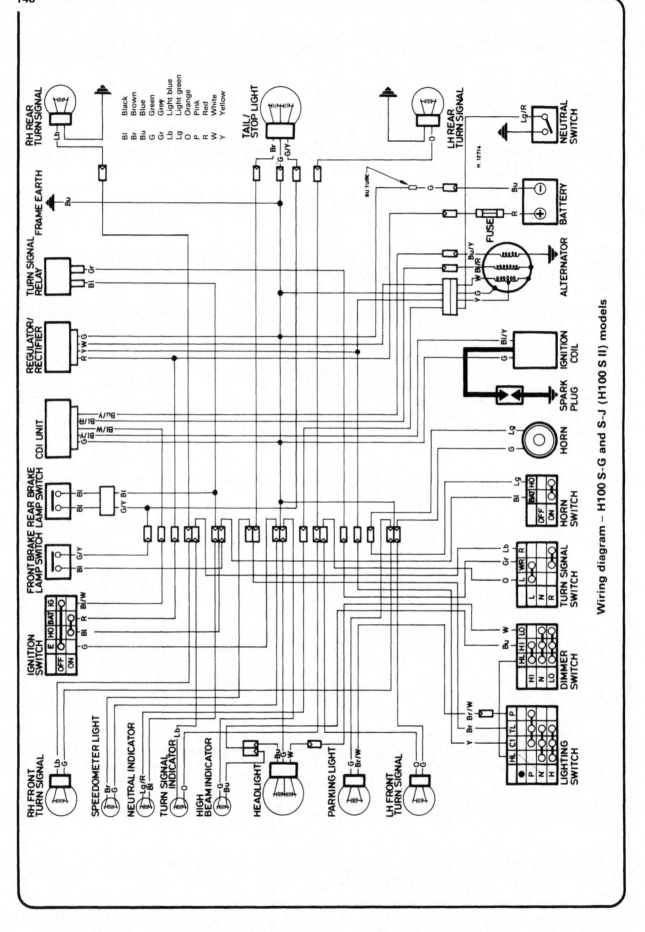

Wiring diagram – H100 S-G and S-J (H100 S II) models

Bl Black
Br Brown
Bu Blue
G   Green
Gr Grey
Lb Light blue
Lg Light green
O   Orange
P   Pink
R   Red
W   White
Y   Yellow

Conversion factors

Length (distance)
	X		=		X		=
Inches (in)	X	25.4	= Millimetres (mm)		X	0.0394	= Inches (in)
Feet (ft)	X	0.305	= Metres (m)		X	3.281	= Feet (ft)
Miles	X	1.609	= Kilometres (km)		X	0.621	= Miles

Volume (capacity)
	X		=		X		=
Cubic inches (cu in; in³)	X	16.387	= Cubic centimetres (cc; cm³)		X	0.061	= Cubic inches (cu in; in³)
Imperial pints (Imp pt)	X	0.568	= Litres (l)		X	1.76	= Imperial pints (Imp pt)
Imperial quarts (Imp qt)	X	1.137	= Litres (l)		X	0.88	= Imperial quarts (Imp qt)
Imperial quarts (Imp qt)	X	1.201	= US quarts (US qt)		X	0.833	= Imperial quarts (Imp qt)
US quarts (US qt)	X	0.946	= Litres (l)		X	1.057	= US quarts (US qt)
Imperial gallons (Imp gal)	X	4.546	= Litres (l)		X	0.22	= Imperial gallons (Imp gal)
Imperial gallons (Imp gal)	X	1.201	= US gallons (US gal)		X	0.833	= Imperial gallons (Imp gal)
US gallons (US gal)	X	3.785	= Litres (l)		X	0.264	= US gallons (US gal)

Mass (weight)
	X		=		X		=
Ounces (oz)	X	28.35	= Grams (g)		X	0.035	= Ounces (oz)
Pounds (lb)	X	0.454	= Kilograms (kg)		X	2.205	= Pounds (lb)

Force
	X		=		X		=
Ounces-force (ozf; oz)	X	0.278	= Newtons (N)		X	3.6	= Ounces-force (ozf; oz)
Pounds-force (lbf; lb)	X	4.448	= Newtons (N)		X	0.225	= Pounds-force (lbf; lb)
Newtons (N)	X	0.1	= Kilograms-force (kgf; kg)		X	9.81	= Newtons (N)

Pressure
	X		=		X		=
Pounds-force per square inch (psi; lbf/in²; lb/in²)	X	0.070	= Kilograms-force per square centimetre (kgf/cm²; kg/cm²)		X	14.223	= Pounds-force per square inch (psi; lbf/in²; lb/in²)
Pounds-force per square inch (psi; lbf/in²; lb/in²)	X	0.068	= Atmospheres (atm)		X	14.696	= Pounds-force per square inch (psi; lbf/in²; lb/in²)
Pounds-force per square inch (psi; lbf/in²; lb/in²)	X	0.069	= Bars		X	14.5	= Pounds-force per square inch (psi; lbf/in²; lb/in²)
Pounds-force per square inch (psi; lbf/in²; lb/in²)	X	6.895	= Kilopascals (kPa)		X	0.145	= Pounds-force per square inch (psi; lbf/in²; lb/in²)
Kilopascals (kPa)	X	0.01	= Kilograms-force per square centimetre (kgf/cm²; kg/cm²)		X	98.1	= Kilopascals (kPa)
Millibar (mbar)	X	100	= Pascals (Pa)		X	0.01	= Millibar (mbar)
Millibar (mbar)	X	0.0145	= Pounds-force per square inch (psi; lbf/in²; lb/in²)		X	68.947	= Millibar (mbar)
Millibar (mbar)	X	0.75	= Millimetres of mercury (mmHg)		X	1.333	= Millibar (mbar)
Millibar (mbar)	X	0.401	= Inches of water (inH₂O)		X	2.491	= Millibar (mbar)
Millimetres of mercury (mmHg)	X	0.535	= Inches of water (inH₂O)		X	1.868	= Millimetres of mercury (mmHg)
Inches of water (inH₂O)	X	0.036	= Pounds-force per square inch (psi; lbf/in²; lb/in²)		X	27.68	= Inches of water (inH₂O)

Torque (moment of force)
	X		=		X		=
Pounds-force inches (lbf in; lb in)	X	1.152	= Kilograms-force centimetre (kgf cm; kg cm)		X	0.868	= Pounds-force inches (lbf in; lb in)
Pounds-force inches (lbf in; lb in)	X	0.113	= Newton metres (Nm)		X	8.85	= Pounds-force inches (lbf in; lb in)
Pounds-force inches (lbf in; lb in)	X	0.083	= Pounds-force feet (lbf ft; lb ft)		X	12	= Pounds-force inches (lbf in; lb in)
Pounds-force feet (lbf ft; lb ft)	X	0.138	= Kilograms-force metres (kgf m; kg m)		X	7.233	= Pounds-force feet (lbf ft; lb ft)
Pounds-force feet (lbf ft; lb ft)	X	1.356	= Newton metres (Nm)		X	0.738	= Pounds-force feet (lbf ft; lb ft)
Newton metres (Nm)	X	0.102	= Kilograms-force metres (kgf m; kg m)		X	9.804	= Newton metres (Nm)

Power
	X		=		X		=
Horsepower (hp)	X	745.7	= Watts (W)		X	0.0013	= Horsepower (hp)

Velocity (speed)
	X		=		X		=
Miles per hour (miles/hr; mph)	X	1.609	= Kilometres per hour (km/hr; kph)		X	0.621	= Miles per hour (miles/hr; mph)

Fuel consumption
	X		=		X		=
Miles per gallon, Imperial (mpg)	X	0.354	= Kilometres per litre (km/l)		X	2.825	= Miles per gallon, Imperial (mpg)
Miles per gallon, US (mpg)	X	0.425	= Kilometres per litre (km/l)		X	2.352	= Miles per gallon, US (mpg)

Temperature
Degrees Fahrenheit = ($^\circ$C x 1.8) + 32

Degrees Celsius (Degrees Centigrade; $^\circ$C) = ($^\circ$F − 32) x 0.56

It is common practice to convert from miles per gallon (mpg) to litres/100 kilometres (l/100km), where mpg (Imperial) x l/100 km = 282 and mpg (US) x l/100 km = 235

Index